NIOBIUM

CHEMICAL PROPERTIES, APPLICATIONS AND ENVIRONMENTAL EFFECTS

CHEMICAL ENGINEERING METHODS AND TECHNOLOGY

Additional books in this series can be found on Nova's website under the Series tab.

Additional e-books in this series can be found on Nova's website under the e-book tab.

MATERIALS SCIENCE AND TECHNOLOGIES

Additional books in this series can be found on Nova's website under the Series tab.

Additional e-books in this series can be found on Nova's website under the e-book tab.

CHEMICAL ENGINEERING METHODS AND TECHNOLOGY

NIOBIUM

CHEMICAL PROPERTIES, APPLICATIONS AND ENVIRONMENTAL EFFECTS

MATHIS SEGERS

AND

THOMAS PEETERS

EDITORS

New York

For permission to use material from this book please contact us:
Telephone 631-231-7269; Fax 631-231-8175
Web Site: http://www.novapublishers.com

NOTICE TO THE READER

Library of Congress Cataloging-in-Publication Data

ISBN: 978-1-62808-257-9

Published by Nova Science Publishers, Inc. † *New York*

CONTENTS

PREFACE

In this book, the authors present current research in the study of the chemical properties, applications and environmental effects of niobium. Topics discussed include the structure and properties of Nb under severe plastic deformation and in high-strength Cu-Nb nanocomposites; the crystal structure of ordered carbide phases and revised sequence of phase transformations associated with the ordering of strongly nonstoichiometric cardides of group V transition metals; niobium-based alloys as hydrogen permeable membrane for hydrogen separation and purification; and electrode processes in anodic oxide films of niobium.

Chapter 1 – One can hardly overestimate the role of niobium in manufacturing various functional materials, as it is one of the main constituents of such important electro-technical materials as high-strength nanocomposites Cu-Nb and multifilamentary Nb_3Sn-based and NbTi superconductors. Besides, nowadays possibilities of obtaining bulk nanostructured materials by various techniques of severe plastic deformation (SPD) are comprehensively studied, and niobium, being a refractory, but on the other hand plastic enough metal, is an ideal object for such studies. The present chapter is a review on evolution of Nb structure at SPD by different techniques and at large plastic deformation by cold drawing of composite wires (in a Cu matrix). The main attention in this chapter is paid to the original studies of the authors (with their coauthors), but the available publications on the topics considered are involved and reviewed as well.

Chapter 2 – Nonstoichiometric carbides are unique entities for exploring atomic ordering. These compounds display a large variety of superstructures with different symmetry. The effect of ordering on all properties of nonstoichiometric carbides is large and comparable with variation of properties in the homogeneity interval of a disordered compound. In the present paper the data on order-disorder phase transformations in strongly nonstoichiometric carbides of Group V transition metals are reviewed. The main ordered phase of these compounds is the superstructure of M_6C_5 type. Taking into consideration recent findings we showed that trigonal (space group $P3_112$) and monoclinic (space groups $C2/m$ and $C2/c$) M_6C_5 superstructures are formed as a result of ordering of nonstoichiometric cubic ($B1$ structure) carbides MC_y of Group V transition d metals. Symmetry analysis of the monoclinic and trigonal M_6C_5 superstructures is performed, $MC_y \rightarrow M_6C_5$ disorder-order transition channels are found, and the distribution functions of carbon atoms in M_6C_5 superstructures are calculated. It is shown that with a decrease in temperature, two physically admissible sequences of transformaions connected with the formation of M_6C_5 phases are possible in MC_y carbides of Group V transition metals. The first sequence MC_y (space group $Fm\bar{3}m$) $\rightarrow$ M_6C_5 (space group $C2/m$) $\rightarrow$ M_6C_5 (space group $C2/c$) includes disorder-order and order-order transformations. The second sequence involves disorder-order transformation MC_y (space group $Fm\bar{3}m$) $\rightarrow$ M_6C_5 (space group $P3_112$) and a polymorphous transformation M_6C_5 (space group $P3_112$) $\rightarrow$ M_6C_5 (space group $C2/c$).

Chapter 3 – Niobium (Nb) is known to exhibit the highest hydrogen permeability among metals so it is one of the most promising metals for the advanced hydrogen permeable membranes for hydrogen separation and purification. However, Nb is very sensitive to hydrogen and when it is exposed to hydrogen atmosphere, a brittle fracture occurs due to severe hydrogen embrittlement. In order to design and develop Nb–based alloys as hydrogen permeable membrane, it is important to improve the resistance to hydrogen embrittlement. For this purpose, it is necessary to investigate the mechanical properties of Nb in hydrogen atmosphere in a quantitative way. From a series of in–$situ$ SP tests, a new phenomenon of the ductile–to–brittle transition of Nb is found, which takes place as a function of hydrogen concentration. The critical hydrogen concentration, DBTC (ductile–to–brittle transition concentration) is found to be about 0.2 (H/Nb), the atomic ratio of hydrogen to Nb. On the bases of these findings, a concept for alloy design of Nb–based hydrogen permeable membrane has been proposed in order to satisfy both high hydrogen permeability and strong resistance to hydrogen embrittlement. Following the concept, Nb–W–Mo ternary alloy has been successively design and developed. The designed alloy possesses about 5 times higher hydrogen permeability than currently used Pd–based alloys without showing any evidence of hydrogen embrittlement.

Chapter 4 – The authors review the following issues: peculiarities of polycrystalline AOP development at niobium in nitrate salt melt at temperatures of Nb_2O_5 recrystallization, the mechanisms of change in structure and properties of niobium pentoxide in the result of electrode processes under cathodic polarization (at the model structure of polycrystalline film), as well as the methods of amorphous electrochromic AOP development and ways of electrochromic processes optimization at such films.

In: Niobium
Editors: M. Segers and Th. Peeters

ISBN: 978-1-62808-257-9
© 2013 Nova Science Publishers, Inc.

Chapter 1

STRUCTURE AND PROPERTIES OF Nb UNDER SEVERE PLASTIC DEFORMATION AND IN HIGH-STRENGTH Cu-Nb NANOCOMPOSITES

V. V. Popov and E. N. Popova*
Institute of Metal Physics, Ural Branch of RAS, Ekaterinburg, Russia

ABSTRACT

One can hardly overestimate the role of niobium in manufacturing various functional materials, as it is one of the main constituents of such important electro-technical materials as high-strength nanocomposites Cu-Nb and multifilamentary Nb_3Sn-based and NbTi superconductors. Besides, nowadays possibilities of obtaining bulk nanostructured materials by various techniques of severe plastic deformation (SPD) are comprehensively studied, and niobium, being a refractory, but on the other hand plastic enough metal, is an ideal object for such studies. The present chapter is a review on evolution of Nb structure at SPD by different techniques and at large plastic deformation by cold drawing of composite wires (in a Cu matrix). The main attention in this chapter is paid to the original studies of the authors (with their coauthors), but the available publications on the topics considered are involved and reviewed as well.

---

* vpopov@imp.uran.ru.

I. INTRODUCTION

Nowadays nanocrystalline materials attract great attention of physicists and materials science researchers all over the world, as they possess unique structure and properties, many of which are of great practical importance [1-3]. According to the terminology accepted by the international journal "Nanostructured materials", crystalline materials are referred to as nanostructured when the sizes of their grains or other structural units are less than 100 nm. As shown by numerous studies, these materials demonstrate unusual mechanical behavior and possess unique properties, such as extremely high strength, superplasticity, enhanced diffusion coefficients compared to ordinary polycrystals, etc. [1-6].

At present there exist a number of techniques for fabrication of bulk nanostructured materials, including various methods of nano-powders synthesis and consolidation. Using these methods, it is possible to obtain materials with grain sizes of several nanometers. However, there are great problems arising in application of these methods, such as high residual porosity, introduction of impurities in powder preparation and consolidation, etc. That is why it is more preferential to use nanostructuring techniques based on severe plastic deformation (SPD) which are intensively developed nowadays [7].

Presently the best developed techniques of SPD for nanostructuring various metals and alloys are equal-channel angular pressing (ECAP) and high-pressure torsion (HPT). However, capabilities of various SPD techniques in grain structure refinement, especially for pure metals, are not well determined yet, and the problem of obtaining uniform nanocrystalline structure with crystallites the sizes of less than 100 nm separated by high-angle boundaries is far from being solved.

It is well known that SPD obtained materials possess non-equilibrium boundaries and high internal stresses, and thermal stability of their structure is low, which hampers fabrication of thermally stable nanostructured materials with specific properties. Particularly, due to the low thermal stability it is very difficult to reveal specific features of the structure and state of grain boundaries in such materials, and it is still a debatable question whether their boundaries are really more non-equilibrium compared to that of ordinary polycrystals. At the same time, understanding of processes occurring at SPD and further annealing is undoubtedly of great scientific and practical importance. Niobium, which is a refractory but at the same time plastic enough metal, is an ideal object for such studies.

Along with Nb in free state, it is of great interest to study it as a constituent of high-strength heavily-deformed Cu-Nb composites which possess unique properties. Particularly, they have the strength higher than 1000 MPa and at the same time electrical conductivity of 70% of that of high-pure copper. This favorable combination of properties make them irreplaceable materials for application as winding wires for high-field pulse magnets with record-breaking magnetic fields of 50-100 T. Such systems are worked out in USA, Japan, European Union countries, and they are required not only for broadening of fundamental investigations, but also for the solution of a number of important practical problems. Using these materials as conductors essentially extends possibilities of fabrication of electronic and electrotechnical devices working in severe conditions such as in aircraft and space industry or in the Far North. In Russia such conductors are worked out at the Bochvar High-Technological Institute of Inorganic Materials [8, 9].

II. EFFECT OF SPD MODE AND REGIMES ON THE Nb STRUCTURE FORMING AT DEFORMATION AND ITS THERMAL STABILITY

As mentioned in the Introduction, various methods of obtaining bulk nanocrystalline (nanostructured) materials based on severe plastic deformation are widely developed now. Contrary to cold rolling or drawing, which cause the refinement, but the structure remains cellular or substructure with low-angle boundaries is formed, SPD enables to create the ultra-fine grained structure with high-angle boundaries.

In early 90[ths] of the last century great interest in the studies of SPD and its effects was initiated mainly by Russian scientists, who demonstrated that applying severe deformations at low homologous temperatures one can obtain nanostructured materials with specific characteristics [10, 11]. Special methods were worked out for SPD realization, the main of which at present are the equal-channel angular pressing (ECAP) [12, 13] and high-pressure torsion (HPT) [14, 15].

The ECAP was worked out as a technique of plastic deformation by shift without changing specimen transverse sections, which makes it possible to deform them repeatedly, and it was further developed as one of the ways of nanostructuring metals and alloys by SPD [10]. At ECAP a billet is repeatedly extruded in a special device through two channels with equal transverse sections, intersecting at the angle φ, which usually is 90°. Additionally, if necessary, deformation can be carried out at enhanced temperatures. Important factors in structure development with billet integrity are the billet orientation and the number of passes through the channels. Dependently on orientation, there are the following routes: orientation is the same at all the passes (route A); after every pass the billet is turned round its longitudinal axis by 90° (route B); after every pass the billet is turned round its axis by 180° (route C). There are also modifications of route B, referred to as B_A and B_C, which differ in the direction of turn between passes. Thus, in route B_A the billet is alternately turned clockwise and counter-clockwise by 90°, and in B_C it is turned only clockwise [16]. These routes differ in shift direction at repeated passes and result in shape modification of spherical cells in the billet bulk at ECAP. As shown in [17], the fraction of high-angle boundaries increases in the following sequence, $B_A \rightarrow C \rightarrow B_C$. According to [18], the expression of the equivalent strain at ECAP with N passes is as follows:

$$\varepsilon_N = \frac{N}{\sqrt{3}}\left[2ctg\left(\frac{\Phi}{2}+\frac{\Psi}{2}\right)+\Psi cosec\left(\frac{\Phi}{2}+\frac{\Psi}{2}\right)\right], \tag{1}$$

where Φ is internal, and Ψ is external angle.

The true logarithmic strain in this case can be calculated as:

$$e = Arsh\frac{\sqrt{3}\cdot\varepsilon_N}{2} = \ln\left\{\left(\sqrt{3}\varepsilon_N/2\right)+\left[\left(\sqrt{3}\varepsilon_N/2\right)^2+1\right]^{1/2}\right\}, \tag{2}$$

where is the equivalent strain for N passes, calculated from (1).

The HPT technique is the development of the idea of Bridgeman anvils. It is the only method which ensures continuous deformation up to extremely high strains without interruption or changing of deformation route. In this technique both initial and final specimens are disk-shaped. An initial specimen is put between anvils and compressed with the pressure of several GPa. Then the lower anvil starts to rotate, and the forces of surface friction make the specimen to deform by shear. The geometric shape of a specimen is such that most of its bulk is deformed in condition of hydrostatic compression under the applied pressure and the pressure from external layers of the specimen. As a result, the specimen does not fracture in spite of very high strain. There are two types of HPT, with open (unconstrained) or closed (constrained) anvils and with anvils provided with cavities [15].

In HPT a specimen undergoes shear strain due to growing angle of rotation of the lower anvil, θ. In this case the shear strain, γ, is calculated as

$$\gamma = \frac{\theta}{h} R, \tag{3}$$

where R and h are the distance from specimen's center and its thickness, respectively. The rotation angle θ relates to the number of revolutions, N, as: $\theta = 2\pi N$.

To compare the shear strain at torsion with the strain at other deformation modes, the former is transformed into the equivalent strain, ε_{eq}, which is equal to:

$$\varepsilon_{eq} = \gamma / \sqrt{3} \tag{4}$$

Capabilities and limitations of ECAP and HPT for nanostructuring and strengthening of various materials compared to traditional techniques of plastic deformation, such as rolling and drawing, are reviewed in [19]. In the opinion of the author of this review, at HPT the impact of high pressure is superimposed with strain gradient hardening. The latter induces the storage of difficult to recover geometrically necessary dislocations, which makes this method the most effective for structure refinement and strengthening.

Out first attempt to obtain nanocrystalline niobium was undertaken in [20,21], where we applied the room temperature HPT. The disks of single crystalline 99.99 % pure Nb, the diameter of 10 mm and thickness of 0.5 mm were deformed at a rate of 0.3 revolutions per minute, by 5 and 10 revolutions, under the pressure of 5.5 GPa. The true strain, e, was calculated as a sum of shear strain, e_{sh}, and compression strain, e_{comp}.

The shear strain was calculated as

$$e_{sh} = \ln\left[(\varphi R) / h\right], \tag{5}$$

where φ is torsion angle in radians, h is specimen thickness in mm, and R is the distance from torsion axis in mm.

The true compression strain was calculated as

$$e_{comp} = \ln(h_0 / h_f) \,, \tag{6}$$

where h_0 and h_f are specimen thicknesses before and after deformation, mm.

The as-calculated true strain in the radius middle is $e = 6.15$ and 7.25 for 5 and 10 revolutions, respectively.

The as-deformed Nb was annealed in vacuum at 400, 500, 600, 700 and 800^0C for 2 h to determine the thermal stability of the structure obtained. The structure was studied by TEM in JEM-200CX microscope, with further treatment of images by computerized program SIAMS-600 which enables to obtain histograms of grain size distributions. Besides, microhardness was measured by a special device attached to optical microscope NEOPHOT-21 and calculated as

$$H = 18540P/C^2 \,, \text{MPa} \tag{7}$$

where P is loading in grams and C is an indentation diagonal in microns, every value of C being calculated as an average for not less than 9 indentations.

As at HPT the strain differs along the radius, growing from the center to edges, we considered and compared the structure in radius middles as representing some averaged picture for every specimen.

The structure of Nb after HPT by 5 revolutions is shown in Figure 1. It is somewhat non-uniform. In most of the areas one can see grains the sizes of 100-120 nm, with wide and curved non-equilibrium boundaries (Figure 1a). Intricate diffraction contrast and moiré pattern inside grains indicate the presence of high internal stresses. That the crystallites observed are grains, and not cells, is evidenced by electron diffraction patterns, in which a large number of reflections, not elongated azimuthally, are located in the Debye rings, i.e. the boundaries are high-angle (insert in Figure 1a). Besides, in dark-field images the contrast changes drastically from one crystallite to another (Figure 1b), which is characteristic of grains separated by high-angle boundaries. However, in some areas the cellular structure is observed, with wider low-angle boundaries, coarser sizes and higher dislocation density.

According to [22], the submicrocrystalline structure formation at SPD starts with cellular structure; then with strain growth the mixed structure is formed, consisting of cells together with microcrystallites separated with closed high-angle boundaries, and, finally, the structure consisting only of microcrystallites is formed. According to this approach, it may be concluded that room temperature HPT of Nb by 5 revolutions results in the structure corresponding mostly to the last stage, i.e., consisting of microcrystallites (grains), as there are only few areas with cellular structure. As seen from Figure 1c, this structure is borderline between nanocrystalline and submicrocrystalline, as the average grain size is 120 nm, and most of grains fall in the range of 80-200 nm.

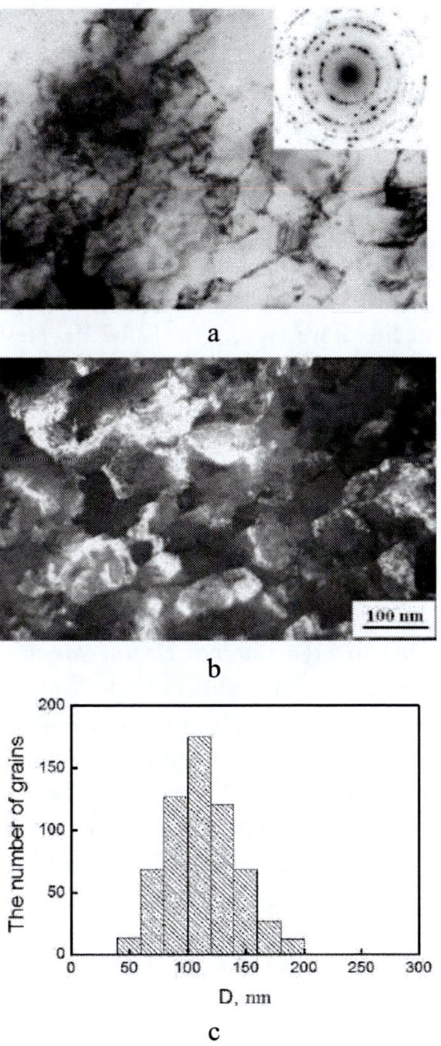

Figure 1. Structure of Nb after room temperature HPT by 5 revs: a – bright-field image; b – dark-field image in (110) Nb reflection; c – histogram of grain size distribution.

For further grain refinement the more severe deformation was carried out, namely, HPT by 10 revolutions. However, no marked changes in the structure were obtained. Though in some areas finer grains than in the previous specimen were observed, the statistical treatment showed that neither the average size, nor the grain size scattering changed. Thus, it can be concluded that in the strain range studied the saturation of structure refinement is observed, and the material resists to further refinement.

The saturation of fragmentation, when further refinement with strain growth is impossible, which hampers obtaining of true nanocrystalline structure in pure metals by SPD, is discussed in [23]. At saturation stage material strengthening is also finished, this behavior being characteristic of all single-phase materials, and it is only the minimal crystallite sizes and the strain at which the saturation stage starts that depend on a material type [24]. The initial process of structure fragmentation and the main parameters affecting grain refinement (temperature, doping, deformation rate and route) are well described in literature (see, e. g, [25-27]). Nevertheless, there is no general opinion regarding the reason of the onset of stable state of microstructure and strengthening with growing strain. For example, in [23] it is suggested that at high homologous temperatures the stable state is caused by the processes similar to dynamic recrystallization, whereas at low homologous temperatures this state is supported by grain boundary migration under stresses.

Besides the saturation of fragmentation, another problem of materials nanostructured by SPD is the low thermal stability of the structure obtained. It is well known that due to high internal elastic stresses and non-equilibrium boundaries the grain growth in such materials starts at temperatures lower than $0.4T_m$ [28]. As the melting temperature of Nb is 2742 K, $0.4T_m$ is 1097 K or $823^{\circ}C$, and $0.3T_m$ is $550^{\circ}C$, and we studied the effect of annealing in the range of $400\text{-}800^{\circ}C$ on the structure of Nb processed by room temperature HPT. This temperature range is of special interest also because the temperatures of diffusion annealing applied in manufacturing of Nb_3Sn-based superconducting composites by solid-sate diffusion between Nb filaments and bronze matrix (the so-called "bronze" process) fall in it [29]. After the annealing at $400^{\circ}C$, 2 h the structure of Nb, processed by 5 and 10 revolutions of HPT, practically does not change. Grain boundaries are still considerably curved, in some grains dislocation density is high, and in some of them the moiré pattern is observed. It may be suggested that at this temperature the recovery only starts, and according to TEM images and histograms of grain size distributions (Figure 2) the structure remains practically the same for both strains studied.

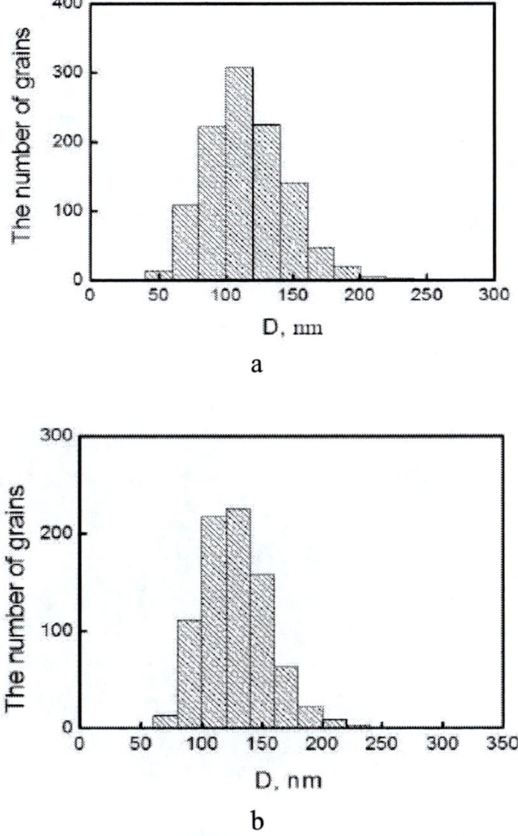

Figure 2. Histograms of grain size distribution in Nb processed by 5 (a) and 10 (b) revolutions of HPT and further annealing at 400°C, 2 h.

At 500-600°C the recovery processes in the specimen, deformed by 5 revolutions, get more pronounced: the dislocation density in grains markedly decreases, many grains are dislocation-free, grain boundaries become thinner and straighter, and some of them form almost regular-shaped polyhedrons, characteristic of the recrystallized state. In some grains the striated contrast is observed, which, in the opinion of some authors [30], indicate the recovery of the structure of grain boundaries (Figure 3a). However, recrystallization, which implies nucleation and *growth* of new grains, does not occur in this case, as the grain sizes in general do not increase.

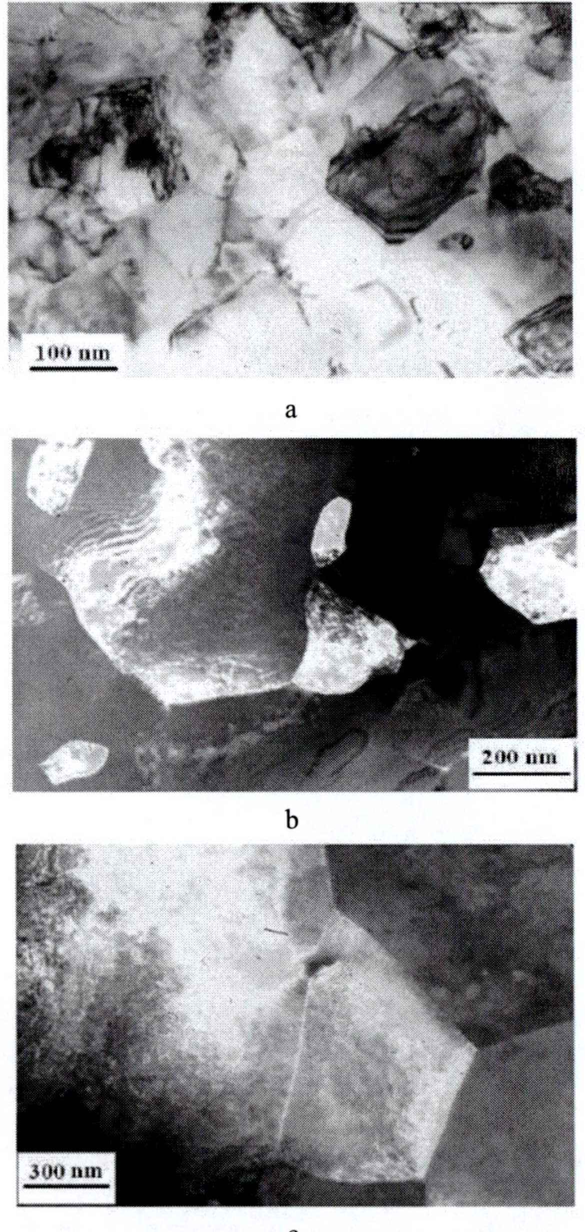

a

b

c

Figure 3. Structure of Nb after HPT and annealing: a - 5 revs + 600°C, 2 h; b - 10 revs + 500°C, 2 h; c - 10 revs + 600°C, 2ч; a – bright-field image; b,c – dark-field images in $(110)_{Nb}$ reflections.

The specimen subjected to HPT by 10 revolutions demonstrates quite a different behavior, i.e. with the annealing temperature growth the effect of strain on the thermal stability of the structure obtained is revealed. After the HPT by 10 revolutions even at as low annealing temperature as 500°C the recrystallization with marked grain growth is observed (Figure 3b). Dislocation-free grains characteristic of the recrystallized states are clearly seen, and the grain size scattering is considerable both within one area and throughout the whole specimen. Grain boundaries get markedly thinner and straighter, which also testifies the recrystallization development.

Thus, it may be concluded that after the higher strain the thermal stability of Nb structure decreases. It is especially obvious after the annealing at higher temperature, 600°C. After such treatment in the specimen deformed by 5 revolutions the structure remains submicrocrystalline, and only the recovery processes occur (Figure 3a), whereas in the more severely strained (by 10 revolutions) specimen grain sizes get an order of magnitude larger, grain size scattering considerably broadens, and the specimen actually transforms into an ordinary polycrystal, the thin straight boundaries in which form polyhedrons (3c). After the annealing at 700°C, 2 h the structure drastically changes in the specimen deformed by 5 revolutions as well. Recrystallization and grain coarsening occur, grain sizes increase by an order of magnitude, and grain boundaries get thin and straight, and form regular-shaped polyhedrons, though in some areas wider dislocation boundaries are also observed. After the highest temperature annealing (800°C, 2 h) the structure of both specimens is practically the same, with coarse grains and wide grain size scattering (Figure 4).

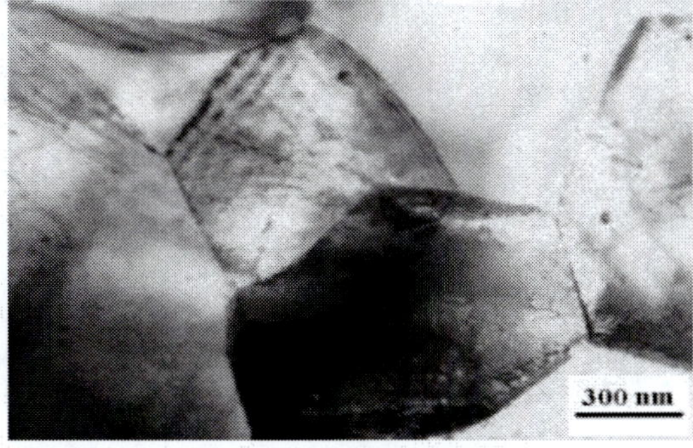

a

Figure 4. (Continued).

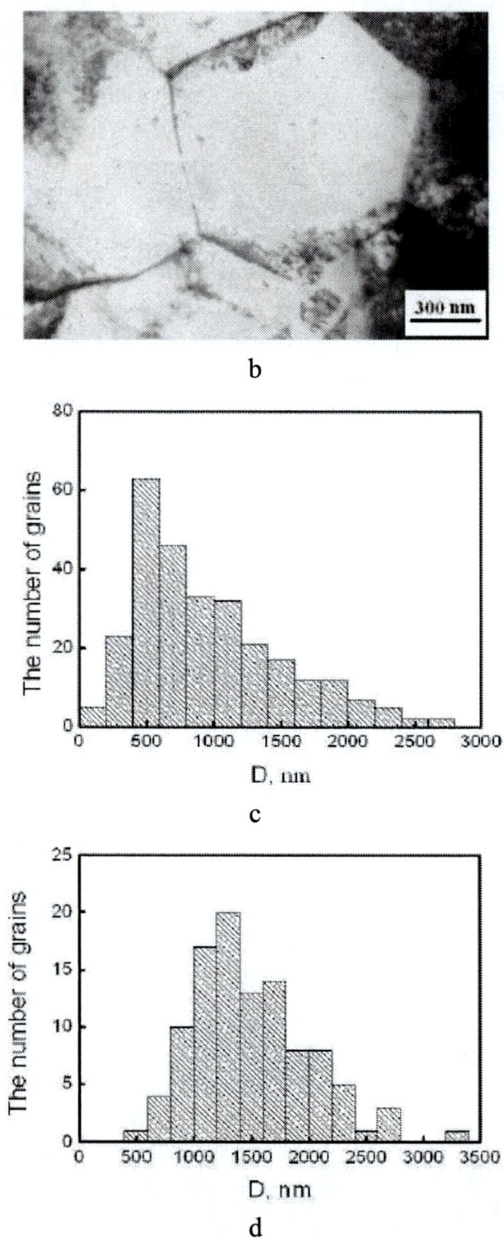

Figure 4. Structure of Nb after HPT by 10 revolutions and annealing (a,b) and histograms of grain size distribution (c,d): a,c – 700°C, 2 h; b,d - 800°C, 2 h.

It was of interest to investigate the microhardness behavior dependently on the strain and the annealing temperature. For convenience of comparison microhardness values are given in Table 1 together with grain size distribution parameters. Microhardness of Nb subjected to HPT by 5 and 10 revolutions is practically equal, about 2520 MPa, which is by a factor of 2.3 higher than that of Nb after multiple pressing and drawing with the same true strain, and it is one more indirect evidence of specific properties of nanostructured materials, particularly, their enhanced strength. A similar result was obtained in [31], where it is shown that tungsten and its alloys in nanostructured state demonstrate microhardness by a factor of 1.35-2 higher than after forging. It should be noted, that higher strain does not result in higher microhardness, as it does not cause further grain refinement.

It is interesting to note that the same high level of microhardness (of about 2500 MPa) is obtained in heavily cold-drawn Cu-Nb composites consisting of Nb filaments (of about 20 volume %) in copper matrix. In that case the enhanced hardness and strength are attributed to specific deformation mechanisms in composites, where Nb filaments are refined to several nm in the transverse sections and acquire the ribbon-like shape, i.e. they also may be considered as nanostructured. Regularities and specific features of nanocrystalline structure formation in such composites are discussed in section IV.

Table 1. Parameters of grain size distributions and microhardness (H) of Nb after HPT by 5 and 10 revolutions and annealing

Treatment	Grain size scattering, nm		D_{av}, nm		RMSD*, nm		H, MPa	
	5 rev.	10 rev.	5 rev.	10 rev.	5 rev.	10 rev.	5 rev.	10 rev.
HPT	45-235	25-230	115	120	29,6	30,1	2530	2520
HPT + 400°C	45-265	60-305	115	130	30,0	28,3	2420	2350
HPT + 500°C	-	120-470	-	240	-	56,5	-	1540
HPT + 600°C	25-220	50-1840	110	1090	28,9	306,2	1800	1150
HPT + 700°C	125-2760	345-2020	970	910	545,7	297,4	1310	1050

*RMSD – root mean square deviation of distribution.

As seen from Table 1, after the annealing at 400°C, 2 h the structure of both specimens remains submicrocrystalline, and their microhardness is practically as high as after the HPT. Annealing at 500°C, 2 h results in the growth of the average grain size and in marked microhardness drop in the more heavily deformed specimen (by 10 revolutions). An appreciable difference in microhardness of the specimens is observed after the annealing at 600°C, 2 h. After 5 revolutions of HPT it decreases only slightly due to the recovery, but it is still quite high, 1800 MPa, whereas after the HPT by 10 revolutions it drastically drops to 1150 MPa, which is not surprising, as the structure of this specimen is less stable, and at this temperature intensive recrystallization proceeds in it.

After the annealing at 700°C the microhardness decreases in both specimens to 1310 and 1050 MPa, respectively, in agreement with structural changes. Nevertheless, the microhardness values are higher than that characteristic of Nb after cold drawing and annealing (800 MPa). A similar result was obtained in the studies of thermal stability of Cr [32]. Annealing of SPD-processed Cr at 700°C causes the uniform coarsening of the structure and microhardness drop, but still the latter remains at a higher level, than in the specimens not subjected to severe plastic deformation. These data demonstrate that at SPD by HPT not only the structure refinement down to nanocrystalline sizes occurs, but a specific stressed state arises, which makes the as-processed material different from ordinary polycrystals.

Thus, HPT of high-pure single-crystalline Nb at room temperature results in the formation of structure borderline between nano- and submicrocrystalline, and we continued to study possibilities of nanostructuring Nb on polycrystalline specimens, applying ECAP, HPT and combination of these methods [33].

The material under study was polycrystalline commercially pure Nb (99.9%) in rods of 10 mm diameter and 60 mm long in the initial state (with grain size of about 20 μm) and after room temperature ECAP by route B_C up to 2, 5 and 16 passes. These samples were kindly given to us by Professor J.G. Sevillano and J. Alkorta (San-Sebastian, Spain), and their treatment is described in detail in [34]. For HPT disks the thickness of 0.5 mm were cut from the middle of these rods and deformed at the rate of 0.3 revolutions per minute under the load of 4.0 GPa. The strain by ECAP was calculated from expression (1), and for every pass the equivalent strain is about 1.15. For 2, 5 and 16 passes studied it is, respectively, 2.30, 5.75 and 18.60. The true strain, e, calculated from (2) is 0.8, 2.2 and 3.4. The structure of specimens was studied by TEM in JEM-200CX and Philips-CM30 microscopes and by SEM in QUANTA-200 microscope equipped by an EDAX device for EBSD analysis. The EBSD analysis was used to determine the fractions of structural elements with different boundary misorientation in the samples after the ECAP. Microhardness was measured as above described for Nb after HPT.

Evolution of structure of polycrystalline Nb under ECAP is shown in Figure 5 which demonstrates the EBSD orientation maps of transverse sections of the specimens. Note that the scale in these orientation maps appreciably differs. Polycrystalline Nb has a typical for such state structure with equiaxed grains the sizes of 20-40 μm, with thin straight boundaries (Figure 5a). After the ECAP by 2 passes (equivalent strain of 2.30) noticeable fragmentation of structure is observed, and in the orientation map one can see areas of fine well-disoriented grains (Figure 5b, lower part of the micrograph) and areas with small misorientation (Figure 5b, upper part). Note that the grain boundaries get less straight than in the initial state.

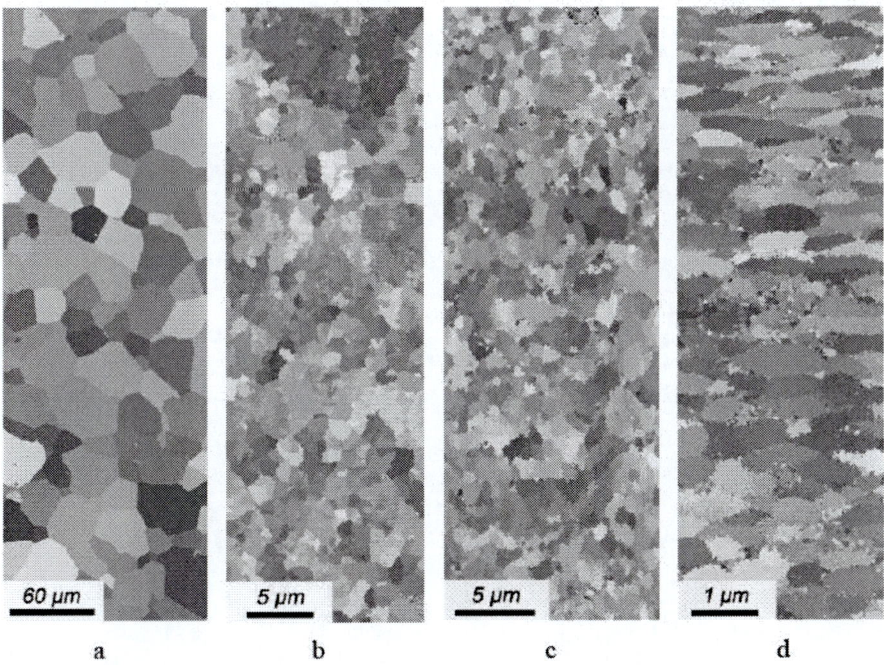

Figure 5. Orientation micrographs of the structure of polycrystalline Nb (a) and Nb after the ECAP to 2.30 (2 passes) (b), 5.75 (5 passes) (c) and 18.60 (16 passes) (d).

Further fragmentation of structure is observed with strain increasing up to 5.75 (5 passes of ECAP) (Figure 5c). As seen from Figure 5c, after 5 passes of ECAP the overall structure is more uniform and there are no wide areas with small misorientation as in the previous case, and the grain boundaries look more distorted. In the specimen subjected to the highest strain of 18.60 (16 passes) the crystallites get finer and some of them acquire a slightly elongated shape (Figure 5d).

Figure 6 demonstrates histograms of grain size distribution in the as-deformed specimens. It is obvious that the initial coarse-grained material acquires the submicrometer-grained structure with growing strain by ECAP. At low and intermediate strain the structure is bimodal. At the strain of 2.30 (2 passes) the sizes of most grains are in the range of 0.5-3 µm, but there are also coarser grains of 5 µm (Figure 6b). With strain growth to 5.75 (5 passes) most of grains fall into the range of 0.4-1.6 µm (Figure 6c). After 16 passes of ECAP further grain refinement is observed, and grain sizes remove to the submicron range of 0.2-0.8 µm, with less grain size scattering (Figure 6d).

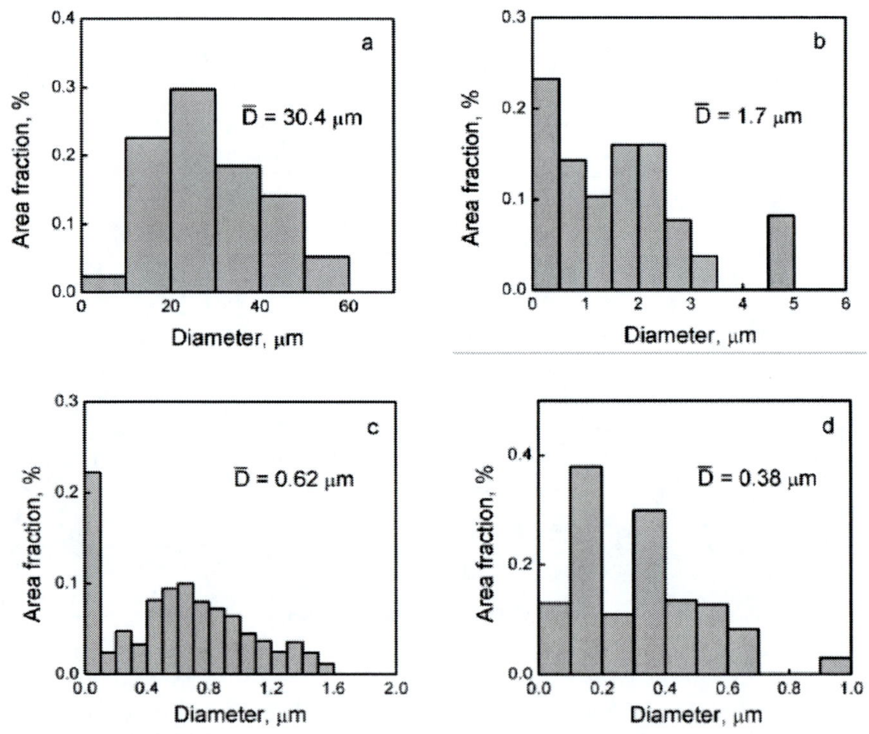

Figure 6. Histograms of grain size distributions in the initial polycrystalline Nb (a) and in Nb after ECAP by 2 (b), 5 (c) and 16 passes (d).

The EBSD analysis allows one to quantitatively estimate the fraction of the structural element boundaries with high and low angles. The transition to the high-angle misorientation is conventionally assumed to occur at 15°. Histograms of misorientation angles are shown in Figure 7. Misorientation angles in polycrystalline coarse-grained state are predominantly in the high-angle range of 45-55 degrees (Figure 7a). After the ECAP to 2.30 the misorientation angle histogram is bimodal, with a noticeable fraction of low-angle boundaries. It is evaluated as 10 %, whereas that of high-angle boundaries is 90 % (Figure 7b). The appearance of the low-angle boundaries testifies that the initial grains are subdivided into dislocation cells with small misorientation. With strain increasing up to 5.75 the fragmentation proceeds, and the relative fraction of low-angle boundaries increases to about 15 % (Figure 7c). At the highest strain degree of 18.60 most of misorientation angles (about 95 %) are high-angle and fall predominantly into the range of 25-55 degrees, though a small fraction of low angles is retained as well (Figure 7d).

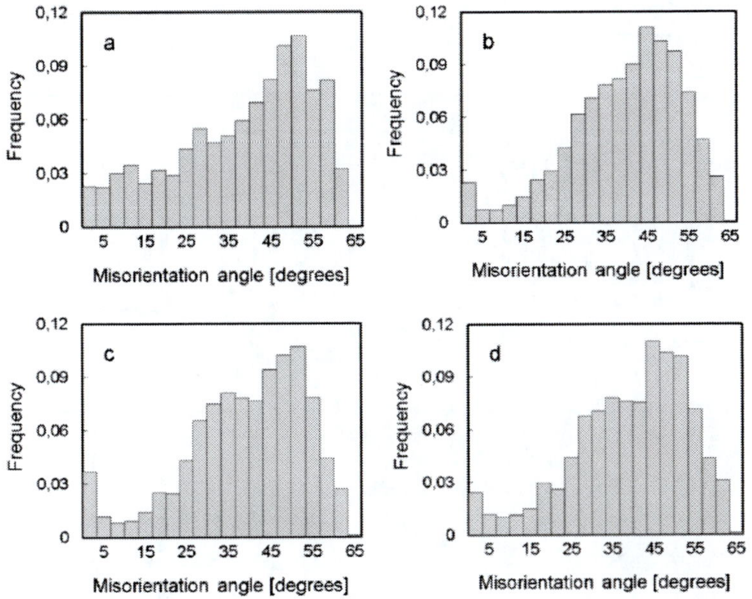

Figure 7. Histograms of misorientation angles in the initial polycrystalline Nb (a) and in Nb after ECAP to 2.30 (b), 5.75 (c) and 18.60 (d).

To examine the fragmentation process and the state of boundaries in more detail, we took fragments of orientation maps in a larger scale and traced the point-to-point misorientation along the AB lines shown in Figure 8. As seen from Figure 8a, after the ECAP to 2.30 coarse fragments with low-angle boundaries are formed inside the initial polycrystalline structure, and along the AB trace one can clearly see a high-angle boundary and several low-angle boundaries (Figure 8b). With strain increasing up to 5.75 the refinement of fragments is obvious (Figure 8c), and along the AB trace there is a number of low-angle boundaries and one high-angle boundary (Figure 8d). At the highest strain of 18.60 only small crystallites with high-angle misorientation are observed (Figure 8 e, f).

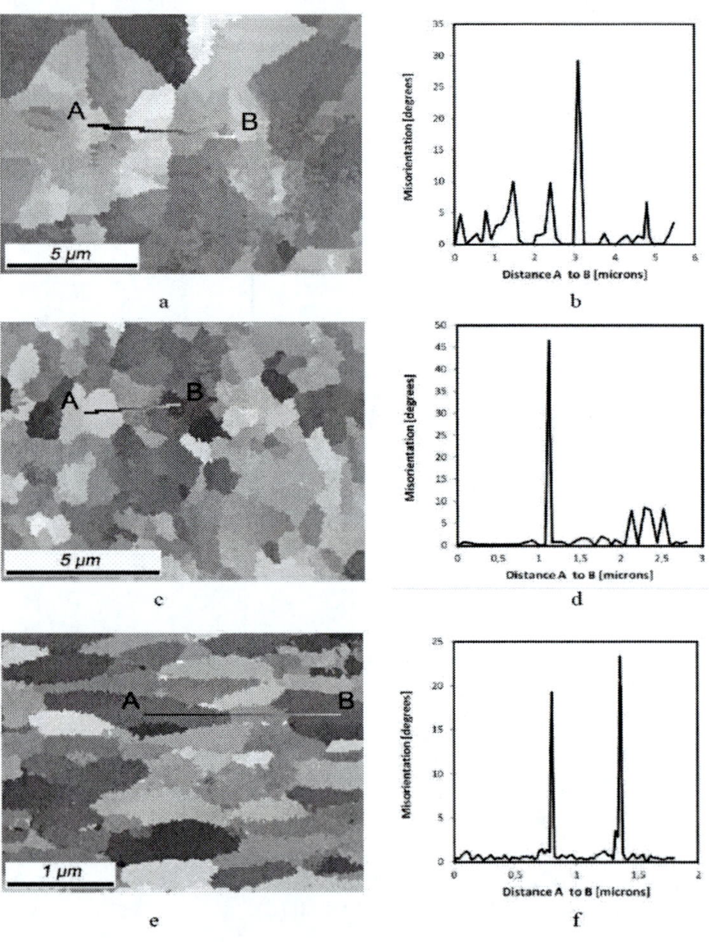

Figure 8. Fragments of orientation maps in a larger scale and point-to-point misorientation along the AB traces shown in the maps: a, b – ECAP to 2.30; c, d - ECAP to 5.75; e, f - ECAP to 18.60.

Thus, fragmentation of structure at ECAP proceeds through gradual increasing of an amount of boundaries, at first low-angle, and then high-angle, and the process is not finished and the saturation state is not achieved even at the highest strain of 18.60 (16 passes) studied. As seen from Figs. 5 and 8, with strain increasing grain boundaries get less straight and more curved testifying their non-equilibrium state. The latter is confirmed by the direct studies of the state of grain boundaries by the emission Mossbauer spectroscopy, the results of which are presented in the next section.

A specific feature of the structure of a specimen subjected to 16 passes of ECAP is an elongated shape of some grains. It contradicts the data by Sevillano [19], who states that periodical non-monotonic passes at ECAP should result in obtaining more equiaxed structures compared to other techniques of plastic deformation. In [35] the structure of Nb after ECAP was studied by FIB-FEG-SEM, and no elongation of grains was revealed. However, we have found this specific feature of the structure both by EBSD and by transmission electron microscopy, the results of which are presented below.

Transmission electron microscopy contrary to SEM is a highly localized method, and if there is a certain non-uniformity of the structure, one can observe areas of various types in one and the same specimen. However, when examining a large number of foils one can reveal the most typical structures of this or that specimen. On the other hand, TEM enables to study specific features of fine structure of crystallites and their interfaces. Note that TEM images were taken from central parts of cross sections of the ECAP-ed specimens.

After the ECAP by 2 passes the dislocation cell structure with remarkable non-uniformity of sizes and shapes of cells is dominating. Some cells are approximately equiaxed, and the others are elongated markedly (Figure 9a). Interfaces between cells have a typical dislocation structure, and in the electron diffraction patterns there are reflections from only one (or more seldom from two) plane. With strain increasing up to 5.75 the structure gets somewhat finer, and subgrain structure with lower dislocation density and thinner subgrain boundaries appears. A remarkable fraction of subgrains the sizes of about 250 nm, of equiaxed or slightly elongated shape are observed, though there are also areas with much coarser crystallites. In the dark-field image shown in Fig 9b one can see small change of contrast between the neighboring crystallites, and in the corresponding electron diffraction pattern there is a tendency of the Debye rings formation testifying crystallites turn relative to each other by several degrees.

Remarkable fragmentation of structure is observed at the ECAP by 16 passes (to 18.60). After such treatment in most areas there are slightly elongated grains and subgrains the sizes of 80-250 nm, with thin boundaries which do not demonstrate pronounced dislocation structure (Figure 9c). In some electron diffraction patterns one can see almost complete Debye rings (Figure 9c, insert), which means that not only subgrain but grain structure has formed as well. However, even in this specimen there are areas with much coarser cells and subgrains, with small contrast changes in dark-field images and high dislocation density.

Thus, the structure obtained by ECAP is in general non-uniform even at the highest strain studied, and the saturation of fragmentation is not reached.

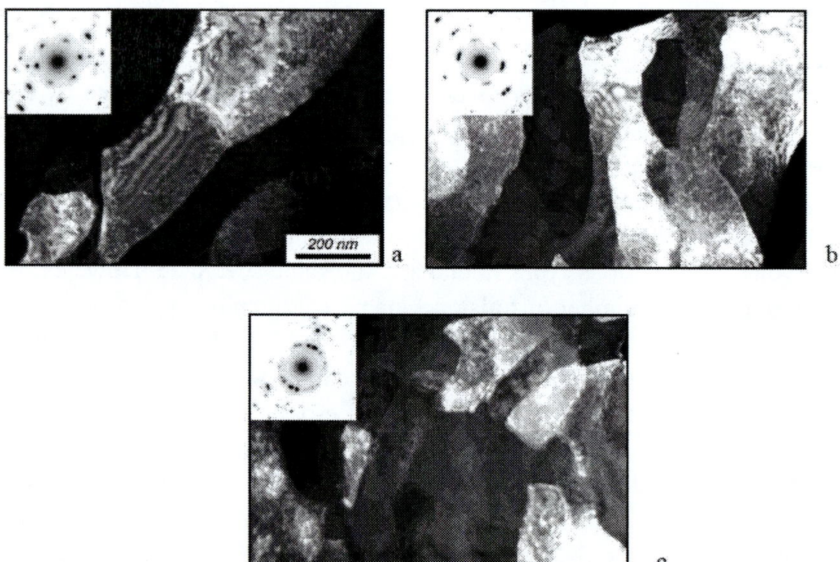

Figure 9. Structure of Nb after the ECAP by 2 (a), 5 (b) and 16 passes (c): dark-field images in $(110)_{Nb}$ reflections and corresponding electron diffraction patterns.

At HPT of polycrystalline Nb by 5 revolutions, as well as single crystalline, the structure reaches the saturation state and remains the borderline between submicro- and nanocrystalline, with crystallite sizes of 100-120 nm, and in some areas not only grains but dislocation cellular structure is observed. It was of interest to reveal whether the preliminary deformation by ECAP imposes any effect on the structure of the HPT deformed Nb, because ECAP results not only in the refinement of structure but it appreciably increases the dislocation density and the level of the internal stresses, and changes the state of grain boundaries which get non-equilibrium.

The structure of specimens deformed by 2 and 5 passes of ECAP and further HPT by 5 revolutions appeared to be practically the same as after the HPT of single- or polycrystalline Nb, and no additional strengthening was achieved. It was expected that at least small effect in grain refinement could be obtained by preliminary ECAP with the highest strain, 16 passes, as this treatment itself resulted in considerable refinement of the structure. The structure of the specimen after the combined deformation by maximal ECAP and HPT is shown in Figure 10. It is seen that in general the structure is not further refined, and it is still in the saturation state with grain sizes of 100-120 nm. The boundaries of some grains are thinner and less curved, and all the electron diffraction patterns are ring-wise, with practically uniform distribution of reflections all over the Debye rings. No areas with cellular or subgrain structure were observed in this case.

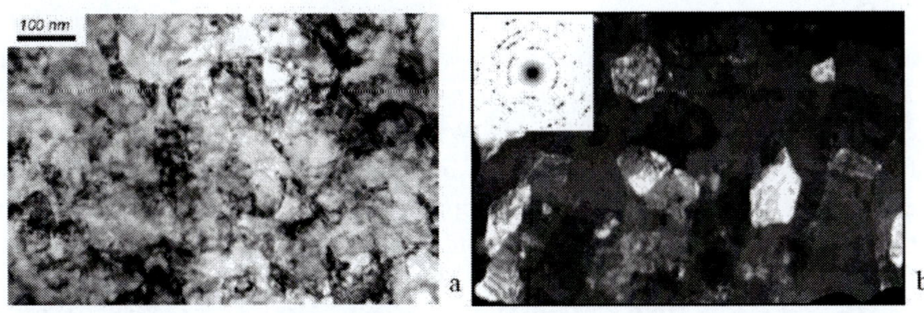

Figure 10. Fine structure of Nb after ECAP by 16 passes + HPT by 5 revs: a– bright-field image; b – dark-field image in $(110)_{Nb}$ reflection and corresponding electron diffraction pattern.

Thus, it may be concluded that at room temperature HPT of Nb by 5 passes the saturation of fragmentation is reached, the structure and microhardness get equal along the specimen radius, and no further refinement occurs. The initial state (single crystalline, polycrystalline or preliminary ECAP) does not affect the structure at the saturation state. The latter is in full agreement with the conclusion made in [23], that HPT of single-phase materials results in one and the same saturation grain size independently of whether the initial structure was coarser or finer.

If we compare the structure after ECAP and after the combination of ECAP and HPT, the difference is obvious. After the HPT the structure is much more dispersed, and the specific curved contrast is observed in grains testifying the high level of internal elastic stresses and higher dislocation density, and it can be concluded that the effect of HPT on the structure is much stronger than that of ECAP. It is confirmed by the measurements of microhardness after various modes of SPD (Table 2).

Table 2. Microhardness (H) of Nb after various modes of SPD

Treatment	H, MPa	Treatment	H, MPa
Polycrystal	800	Polycrystal + HPT by 5 revolutions	3100
ECAP to 2.30 (2 passes)	1670	ECAP by 2 passes + HPT by 5 revs	3170
ECAP to 5.75 (5 passes)	1880	ECAP by 5 passes + HPT by 5 revs	3190
ECAP to18.60 (16 passes)	2180	ECAP by 16 passes + HPT by 5 revs	3150

As seen from this table, ECAP by only 2 passes results in a drastic (by a factor of 2) increase of microhardness compared to the initial polycrystalline state. The authors of [34] explain such effect by the formation of additional geometrically necessary dislocations. With growing strain (increasing number of passes) microhardness increases, but much slower, as the density of geometrical necessary dislocations approaches its saturation. An analogous effect was observed also for heavily cold-drawn Nb [36].

ECAP even with the highest strain (16 passes) gives much lower values of microhardness than HPT by 5 revolutions. In the opinion of Selillano [19], at HPT the effect of high pressure on the evolution of structure is superimposed with strain gradient hardening, which causes the storage of high density of geometrically necessary dislocations and gives additional strengthening due to more effective formation of nanodomains.

As demonstrated above, the HPT has a stronger impact on the structure than ECAP, and the more intensive strengthening in case of the former is quite expectable. Combination of ECAP and HPT gives only a small increment of microhardness, and in this case also some saturation is observed and even a slight decreasing of microhardness at the highest strain by preliminary ECAP. A similar effect is described above for HPT of single crystalline Nb, namely, with strain increasing from 5 to 10 revolutions there was no additional refinement or strengthening.

The saturation of fragmentation at SPD was reported by many authors (see, e.g., [23, 27, 37-39]), but the reasons for this effect were explained differently. Most probably, in case of intermediate and high temperatures the main reason is the dynamic recrystallization. However, in case of low homologous temperatures it is unlikely, because recrystallization, including dynamic, is a diffusion process, and at low temperatures it is suppressed. In [40] it is suggested that dynamic recrystallization is possible at room temperature due to high internal stresses which can accelerate diffusion. Another suggestion is made in [23], that the processes causing the saturation stage differ from the dynamic recrystallization, though they have much in common. The main argument against the dynamic recrystallization is that in most cases the grain sizes after SPD are of the same order of magnitude or even smaller than typical nuclei of recrystallization, which makes the latter hardly possible. Therefore, it is concluded that the saturation stage is caused by grain boundary migration under high stresses.

In our opinion, in case of refractory metals, particularly, niobium, the dynamic recrystallization at room temperature SPD is highly improbable, as for Nb the homologous temperature in this case is only 0.11; and it is grain boundary migration under high stresses that causes the saturation of fragmentation.

One of the ways to suppress grain boundary migration and achieve more intensive refinement is to decrease the deformation temperature. That is why we applied HPT at cryogenic temperature (in liquid nitrogen) for nanostructuring Nb [41]. Disks of 99.99% pure single-crystalline niobium the diameter of 6 mm and the thickness of 0.5 mm were deformed at the rate of 0.3 revolutions per minute under the load of 10 GPa from 0.25 to 5 revolutions. The temperature of specimens was 80 K.

Since at high-pressure torsion the strain is non-uniformly distributed along the specimen radius, growing from the center to periphery, the structure also refines non-uniformly, and microhardness differs along the radius as well. On the other hand, many experimental data, including ours, indicate, that this non-uniformity of structure and properties, particularly, microhardness, level off with strain growth. It is illustrated by Figure 11, in which microhardness of specimens deformed in liquid nitrogen with different strain is plotted versus the specimen radii. At low strains (low angles of anvil revolution) microhardness in the center and periphery differs noticeably, whereas at HPT by 3 and 5 revolutions it is practically the same throughout the cross section, and at 3 revolutions the saturation stage is reached, after which there is no further increase of microhardness with strain growth.

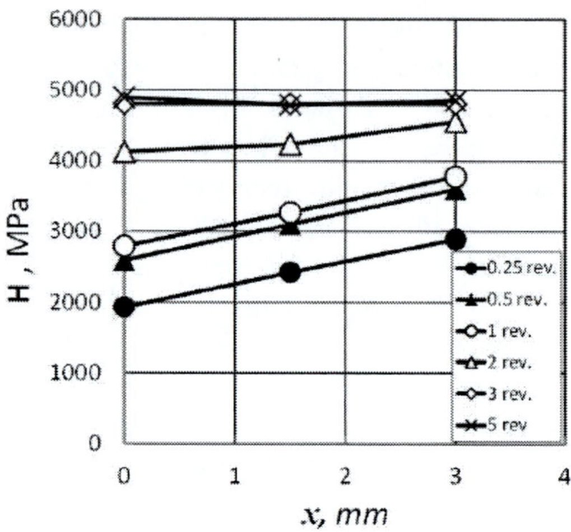

Figure 11. Microhardness of Nb along the radius at different strains by HPT in liquid nitrogen.

It should be noted that the microhardness of this specimen in the saturation stage reaches the record-breaking for Nb value of 4800 MPa, and its value is practically the same along the specimen's radius contrary to the specimens with lower strains. The value obtained is about twice as high as for niobium nanostructured at room temperature or for high-strength heavily-drawn Cu-Nb composites, in which anomalously high strength is due to nano-scaled sizes of Nb filaments and interspaces between them in copper matrix (the latter are discussed in detail in section IV of this chapter).

Evolution of Nb structure at HPT in liquid nitrogen may be traced either in one specimen, moving from the center to periphery, or comparing the structure in the radius middle of specimens deformed with growing strain from 0.25 to 3 revolutions, i. e. to saturation state. As mentioned above, at HPT grain refinement proceeds through three main stages, the first of which is the formation of the dislocation cell structure. With strain growth dislocation density increases, and dislocations are not randomly stored in the structure, but concentrate mainly in cell boundaries. At the second stage, with further strain increasing, a mixed structure is formed, containing both cells with somewhat smaller sizes and subgrains with growing misorientation angles between them. And, finally, the third stage is characterized by uniform nanostructure with crystallite sizes of about 100 nm and mainly high-angle boundaries between them.

After 1 revolution HPT in liquid nitrogen all three stages of Nb structure evolution are observed (Figure 12). In the specimen center dislocation cell structure with high dislocation density and markedly non-uniform dislocation distribution, with coarse cells the size of 0.5-0.9 μm is observed (Figure 12a). In the electron diffraction patterns the reflections of only one plane are present. At the radius middle mixed structure is observed, and in the electron diffraction patterns several spots located near each other in the Debye rings indicate the formation of low-angle boundaries (Figure 12b). And, finally, on the specimen's edge after this treatment the nanocrystalline structure is clearly seen, and the electron diffraction patterns are ring-wise, with randomly located bright spots all over the Debye rings (Figure 12c).

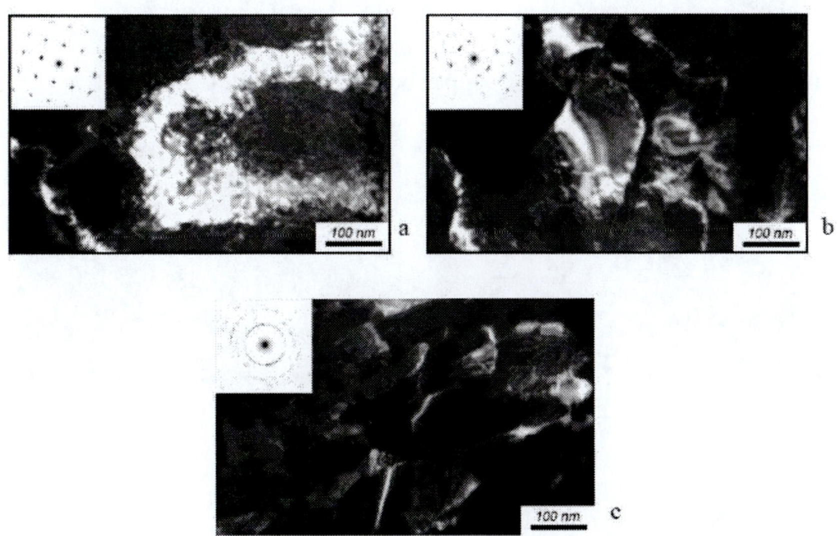

Figure 12. Structure of Nb after 1 revolution HPT at 80K: a – in the specimen's center; b – in the radius middle; c – at the periphery; dark-field images in $(110)_{Nb}$ reflections and electron diffraction patterns.

With increasing strain at 80 K up to 3 revolutions of anvils ($e = 7.1$ in the radius middle) the structure changes drastically. It becomes nanocrystalline throughout the specimen's cross-section, and both the average grain size (75 nm) and the grain size scattering (30-180 nm) are considerably lower than that attained at room temperature HPT. It is interesting to note that in this case the structure is quite uniform along the specimen's radius. A specific curved contrast characteristic of HPT treated materials is observed in grain bulk, and all the electron diffraction patterns are ring-wise, that is, they consist of bright point-wise reflections, not elongated in azimuth directions and located along entire Debye rings (Figure 13: here and further the image taken at the radius middle is shown). The latter reliably demonstrate that most of crystallite boundaries are high-angle, with random misorientation between grains.

It is important to note that the as-obtained structure appeared to be stable at room temperature, and its 6 months ageing did not result in either grain growth, or microhardness drop. Consequently, one can conclude that niobium nanostructured at cryogenic temperature does not undergo recrystallization at ambient temperature contrary to copper, in which we also managed to refine the structure up to the nanocrystalline state by HPT in liquid nitrogen, but it appeared to be extremely unstable and undergone recrystallization during a very short period, from several minutes to several days dependently on the strain degree [42]. Actually, that is not surprising taking into account that room temperature is $0.22T_m$ for copper but only of about $0.1T_m$ for Nb.

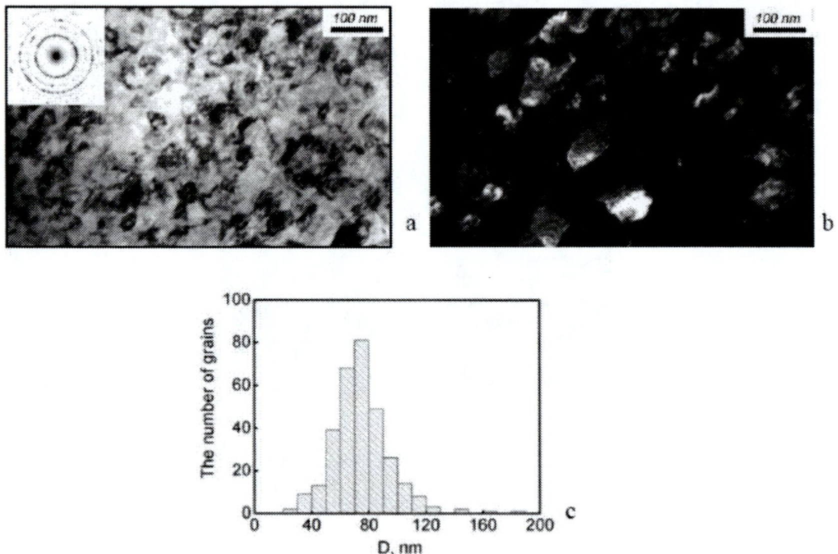

Figure 13. Structure of Nb after HPT by 3 revolutions at 80 K: a – bright-field image with electron diffraction pattern; b – dark-field image in $(110)_{Nb}$ reflection; c – histogram of grain size distribution.

It should be noted that the evolution of structure and microhardness growth at low-temperature HPT of Nb differ from the data obtained for similar treatment of Fe [43] and nickel [44]. In these materials at cryogenic temperatures before the formation of nanocrystalline structure more non-uniform structures than at room temperature HPT are formed. In Ni and Fe the band-wise structures appear, containing twins and deformation localization bands. It is due to the decreasing of dislocation mobility in them and hence to the appearance of deformation localization and twinning modes. In Nb the deformation temperature decrease does not result in band-wise structure formation, which can be explained by smaller change in its homologous temperature and by its essentially higher stacking fault energy. Thus, the decrease of dislocation mobility in Nb at low temperature of deformation at high pressure is not as crucial and does not have such a pronounced effect as in other metals. The nanocrystalline structure formation in Nb at cryogenic temperature proceeds through an appearance and evolution of cellular structure, as is the case at room temperature.

As demonstrated above, the thermal stability of structure of Nb processed by room temperature HPT decreases with the growth of strain, and after the HPT by 10 revolutions intensive grain growth is observed at 600°C annealing, whereas after 5 revolutions only recovery is observed at this annealing temperature.

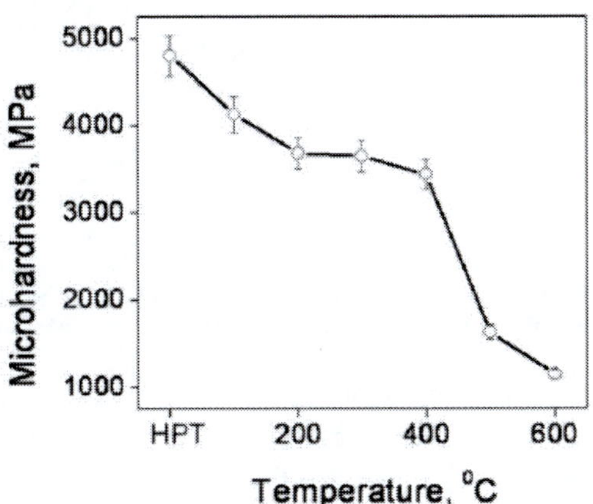

Figure 14. Microhardness of Nb after HPT by 3 revs at 80 K and further annealing.

The thermal stability of Nb structure, obtained by HPT in liquid nitrogen, was studied on the specimen deformed by 3 revolutions. As mentioned above, this structure appeared to be quite stable at room temperature, and either grain growth, or microhardness drop were not found in it during 6 months ageing.

Figure 14 demonstrates microhardness versus the annealing temperature for the specimen after HPT by 3 revolutions at 80 K, and in Figure 15 the histograms of grain size distributions are shown. As seen from these figures, after the annealing at 100-200°C microhardness decreases slightly, while the crystallite sizes practically do not change. At 100°C the microhardness decreasing is most probably due to recovery, and at 200°C some grain growth is added, as the average grain size gets larger.

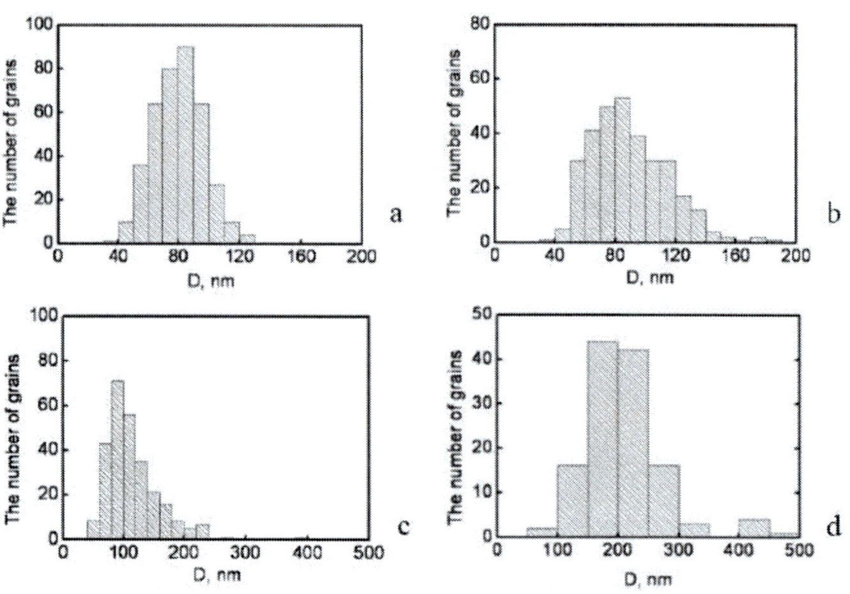

Figure 15. Histograms of grain size distributions after HPT of Nb in liquid nitrogen by 3 revs and further annealing at 100°C (a); 200°C (b); 300°C (c) and 400°C (d).

At higher annealing temperature (300°C) the grain boundaries get straighter, and the specific curved contrast inside grains is not as pronounced as at lower temperatures. Along with fine grains the sizes of about 100 nm, there appear relatively coarse grains, the sizes of up to 300 nm, the boundaries of which intersect under the angles of 120°, which is characteristic of recrystallized grains (Figure 16a). Thus, after such treatment the grain growth starts at as low as 300°C. However, the relative fraction of recrystallized grains at this annealing temperature is small, and microhardness does not drop markedly compared to the 200°C annealing. Nevertheless, at this annealing temperature the structure gets borderline between nano- and submicrocrystalline.

The further drop of microhardness is observed at 400°C annealing, and at this temperature the fraction of recrystallized grains increases, and the average grain size reaches 210 nm, i.e., the structure is submicrocrystalline (Figure 16 b). Intensive grain growth is observed at 500°C (Figure 16c). Grain sizes reach 1 µm and even more. Microhardness drops down to the values found after the annealing at 500-600°C in specimens processed by 10 revolutions of room temperature HPT. Thus, beginning from 500°C the structure is completely recrystallized throughout the specimen considered.

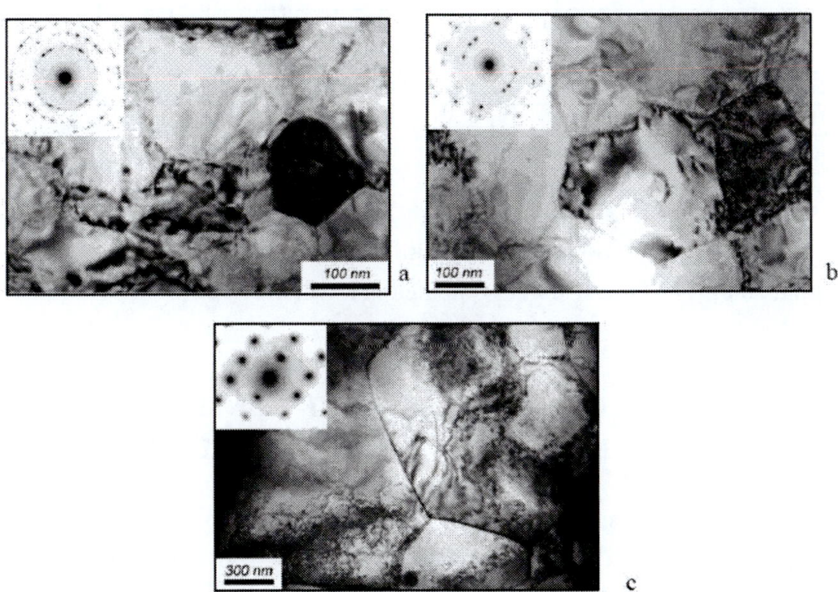

Figure 16. Structure of Nb after HPT by 3 revs at 80 K and annealing at 300°C (a); 400°C (b) and 500°C (c).

Grain sizes and microhardness of specimens annealed after various regimes of HPT are compared in Table 3. It is seen that in specimens deformed by 5 revolutions at room temperature the structure is stable to 600°C. The strain growth up to 10 revolutions (at room temperature) does not result in grain refinement, but the structure gets less stable, and intensive grain growth is observed at 500°C. In specimens, deformed in liquid nitrogen, the grain growth starts at as low as 400°C.

Table 3. Average grain sizes (D_{av}) and microhardness (H) of specimens studied[*]

Treatment	D_{av}, nm			H, MPa		
	5 revs, $T = 20^0C$	10 revs, $T = 20^0C$	3 revs, $T = -193^0C$	5 revs $T = 20^0C$	10 revs $T = 20^0C$	3 revs $T = -193^0C$
HPT	115	120	75	2530	2510	4800
HPT + 100^0C	-	-	80	-	-	4130
HPT + 200^0C	-	-	90	-	-	3690
HPT + 300^0C	-	-	115	-	-	3600
HPT + 400^0C	115	130	210	2420	2350	3440
HPT + 500^0C	-	245	>1000	-	1540	1630
HPT + 600^0C	110	~1000	>1000	1800	1150	1150
HPT + 700^0C	~1000	~1000	>1000	1310	1050	-

[*] the data are given for radius middle.

Thus, decreasing the deformation temperature, it is possible, on the one hand, to obtain the true nanocrystalline state, but, on the other hand, the thermal stability of the structure obtained is lower.

III. SPECIFIC FEATURES OF THE STATE OF GRAIN BOUNDARIES IN THE NANOSTRUCTURED NB

Most of materials, especially metals and alloys, are used in polycrystalline state, the integral part of which are grain boundaries, appreciably affecting many of their practically important properties. The well-established concepts of internal arrangement and properties of grain boundaries have been developed only recently, and they are generalized in a number of monographs and reviews (see, e.g., [45, 46]).

A grain boundary is determined as an interface between two perfect single-phase crystals (or grains) with different crystallographic orientation, which are in contact with each other. According to modern concepts, the grain boundary proper is an area where coordination of atoms differs from that in the perfect lattice, and in metals the width of this area does not exceed 2-3 lattice parameters. Due to the formation of specific structure in the vicinity of grains contact, differing from that of the crystal bulk, the properties of grain boundaries must also be different, and hence sometimes grain boundaries are considered as an independent phase separated from the volume.

The state of grain boundaries to a great degree determine properties of metals, particularly, such processes as grain-boundary sliding, superplasticity, intergranular fracture, etc., the role of boundaries being especially high in nanostructured materials, where their fraction may reach several tens of percents.

Till early 80^{ths} the "equilibrium" grain boundaries were mainly studied, though it is only a conditional conception, because in principle every boundary is non-equilibrium because it is a lattice disruption. Nowadays most of studies deal with the so-called "non-equilibrium" boundaries, which are characteristic of submicro- and nanocrystalline materials obtained by severe plastic deformation. Specific features of boundaries in the non-equilibrium state change the deformation and recrystallization behavior of materials and affect interaction of boundaries with impurities. The term "non-equilibrium" boundaries is widely used in literature since the end of the past century, and very different states of boundaries are referred to as non-equilibrium, e.g., boundaries with non-equilibrium concentration of point defects, boundaries with lattice and extrinsic dislocations, etc. [47-51].

Equilibrium state of grain boundary structure is characterized with minimal free energy at given crystal-geometrical parameters and external conditions (temperature, pressure, concentration of impurities). If only grain-boundary defects of dislocation origin are considered, the above definition of equilibrium of boundaries is equivalent to a requirement that the boundary does not have long-range fields of stresses and its structure is stable. Equilibrium structure of a grain boundary is, as a rule, characterized with periodical distribution of structural elements, dislocation dipoles or grain-boundary dislocations. Consequently, interaction forces between its elements are balanced, and there is no strain in crystals separated with equilibrium boundary. Equilibrium state of grain boundaries can be obtained by prolonged annealing at high enough temperatures. The non-equilibrium state of grain boundary structure is characterized with an enhanced energy, and it is caused by the presence of defects in the boundary structure. Grain boundaries are effective barriers for lattice dislocation movement, and that is why, at external impacts, e.g., at plastic deformation or recrystallization annealing the lattice dislocations are captured by boundaries and stored. The dislocation capture by a boundary results in the formation of a system of primary grain-boundary dislocations which disturbs the equilibrium condition. In this state the excess energy and the long-range stresses are practically completely determined by external dislocations, their type, density and distribution.

Various methods are used to determine the non-equilibrium state of grain boundaries. For example, one can judge on such state by the presence of long-range elastic stresses in near-boundary areas, and this effect may be revealed by modification of contrast in electron-microscopic images. In the very first studies of nanocrystalline materials obtained by SPD the specific appearance of grain boundaries was observed compared to that in ordinary annealed polycrystals, particularly, a marked broadening of thickness extinction contours [11]. The latter is due to high level of internal stresses and distortions near grain boundaries. Certain opportunities for observation of non-equilibrium state of grain boundaries are given by high-resolution electron microscopy. By this method one can reveal high density of various defects: steps, facets, dislocations, the presence of which cause elastic distortions near boundaries [52]. However, quantitative estimation of properties of such boundaries is extremely difficult and requires special investigation techniques, such as, e.g., Mössbauer (nuclear gamma-resonance- NGR) spectroscopy.

Capabilities of the Mössbauer spectroscopy for grain boundary and interface studies in various materials are reviewed in [53]. According to numerous resent publications, as well as our data [54, 55], the most effective method to study grain boundaries in polycrystalline materials is the emission Mössbauer spectroscopy, in which a specimen serves a source of gamma-quanta. This method of investigation of grain boundaries and adjacent areas, worked out in [56], is as follows. Atoms of a Mössbauer radioisotope capable to give spectral information are inserted by diffusion in the material under study at such annealing temperatures, when they localize predominantly in grain boundaries. The registered radiation gives information on the structure and properties of those states, in which the Mossbauer atoms appear to locate. In [57-61] and a number of other studies this method was successfully applied for investigation of grain boundaries in polycrystals of a number of transition and noble metals. In [62, 63] it was used for grain boundary studies in polycrystalline and nanocrystalline niobium. Differences in the parameters of the grain-boundary Mossbauer spectra of poly- and nanocrystalline Nb allow to judge on differences of the state of grain boindaries in these materials.

Specimens of polycrystalline Nb were obtained by rolling of Nb single crystals to 98 % with further annealing at 1073 K (800°C). After such treatment the structure consists of equiaxed grains the sizes of about 50 μm. A relatively high temperature of recrystallization annealing ensures formation of equilibrium grain boundaries and excludes the possibility of their migration at preparation of specimens-sources and further annealing. The nanostructured Nb was obtained by room temperature HPT by 5 revolutions, as described above. In this case one cannot completely exclude the grain boundary migration at annealing. However, as demonstrated above, in the as-obtained specimens the grain growth is not observed at temperature up to 600°C, and hence the grain boundary migration may not be taken into account.

To obtain specimens-sources for the emission Mössbauer studies the ^{119m}Sn radionuclide was electrolytic deposited on one of the surfaces of Nb samples the area of about 1 cm^2. To insert ^{119m}Sn atomic probes in grain boundaries, polycrystalline Nb was annealed at 680K (0.25T$_m$) and nanocrystalline Nb at 574K (0.21T$_m$) for 7 hours. After the annealing the residual non-diffused ^{119m}Sn isotope was removed from the surface together with a thin Nb layer. The removal of the surface layer is also required because at electrolytic deposition and further annealing new phases could form, which hampers the spectra expansion and their interpretation. For instance, the authors of [64] came across this problem in their emission Mössbauer studies of grain boundaries in polycrystalline copper.

Mössbauer spectra were taken from the as-obtained specimens-sources, then annealing at higher temperatures was carried out, and after every annealing a spectrum was taken. In such experimental procedure the Mössbauer atoms remained in the specimen after the first annealing (and the removal of the surface layer), at further annealing just redistribute between grain boundaries and surrounding matrix, at constant total amount of the radionuclide.

Emission NGR spectra of ^{119m}Sn in poly- and nanocrystalline Nb are shown in Figure 17. It is obvious that they are qualitatively similar and posses two components in both cases. The similar result was obtained for all polycrystalline metals studied previously [53-61], and it has been shown that one spectrum component (component 1) is given by the Mössbauer atoms localized in grain boundaries, while the other one (component 2) is formed by the atoms located in near-boundary areas. With temperature growth, the contribution of component 1 decreases, while that of component 2 increases. Similar to the previous investigations, we have found component 2 in the spectrum even at the lowest temperatures of the diffusion annealing (0.21T$_m$ for nano- and 0.25T$_m$ for polycrystalline Nb).

To analyze the results obtained we used the recently suggested model of grain-boundary diffusion [65, 66], which can be considered as the classical Fisher's model specification [67]. According to this model, the local equilibrium is established at an interface between a grain boundary and surrounding matrix. At low temperatures, when the volume diffusion is suppressed, the impurity atoms diffusing along grain boundaries penetrate only into monatomic near-boundary areas, which acquire equilibrium composition. Equilibrium concentration in these near-boundary layers is established due to a reaction at an interface between grain boundary and surrounding matrix. This reaction is assumed to proceed much faster than volume diffusion, and the diffusant penetrates only into near-boundary areas, but not into crystallite bulk. Its penetration into farther matrix areas results from volume diffusion.

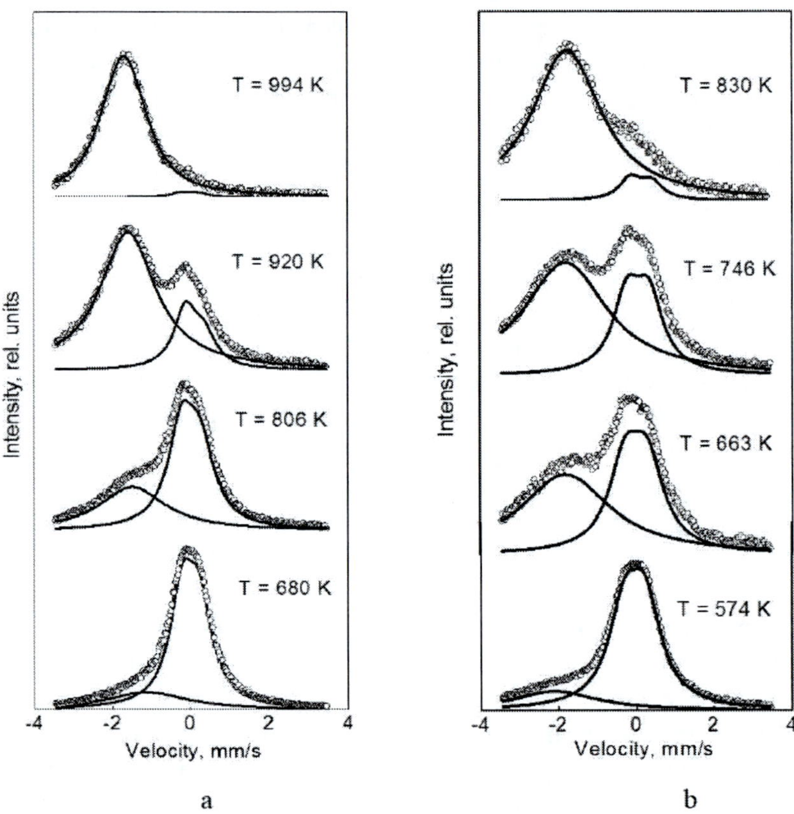

Figure 17. Mössbauer ^{119m}Sn spectra in polycrystalline (a) and nanocrystalline (b) niobium. Annealing temperatures are indicated in the figures.

Figure 18 demonstrates temperature dependences of relative intensities of spectral lines for poly- and nanocrystalline Nb. Two sections are distinctly seen in the temperature dependences of spectral line relative intensities for polycrystalline Nb. At low temperatures (up to about $0.3T_m$) the intensities only slightly depend on the diffusion annealing temperature. From the viewpoint of the model considering the existence of equilibrium composition near-boundary layers, this section corresponds to the redistribution of the diffusing impurity between a grain-boundary and near-boundary layers due to the temperature dependence of the segregation factor. The high-temperature section, in which the temperature dependence of spectral line relative intensities is more drastic, corresponds to the beginning of the impurity atoms diffusion into crystallites bulk.

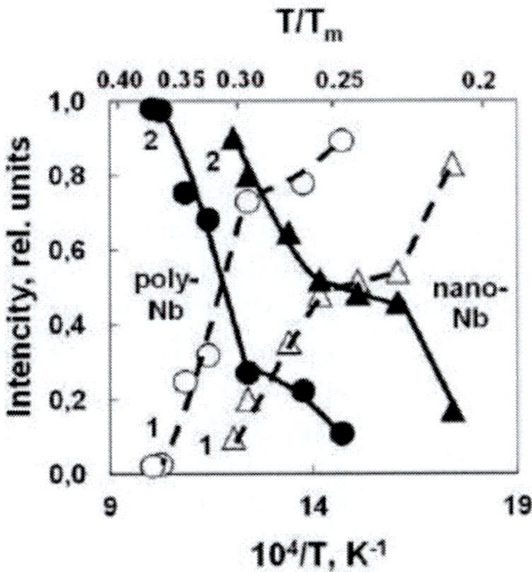

Figure 18. Temperature dependences of relative intensities of ^{119m}Sn spectral lines in poly- and nanocrystalline Nb. Numbers 1 and 2 denote the spectrum components.

Temperature dependences of spectral line relative intensities for nanocrystalline Nb are somewhat different and shifted into the range of lower temperatures. For the lowest annealing temperature, 574K (0.21T$_m$), the intensities ratio is approximately the same as for the low temperature range of polycrystalline Nb. At temperature growth up to 623K (0.23T$_m$) the grain-boundary line (component 1) intensity abruptly decreases, and that of the volume line (component 2) increases. There may be two reasons to explain such a drastic change in the spectral line intensities ratio with small temperature growth. Firstly, it is possible that 574K is a too low temperature for equilibrium to establish at an interface of a boundary and near-boundary layers. Secondly, it can't be excluded that even at this temperature the low-distant migration of some mostly non-equilibrium sections of grain boundaries can already occur. This migration is not observed in structural studies, as it does not exceed several inter-atomic distances, but it is enough to transfer an appreciable amount of the Mössbauer atoms from a grain-boundary into the near-boundary areas of the regular lattice.

In the temperature interval of 623-706K (0.23-0.26T_m) spectral line intensities relatively weakly depend on temperature, the same as in case of polycrystalline Nb. Obviously, in this range the ^{119m}Sn isotope atoms mainly redistribute between grain-boundaries and the surrounding matrix.

At higher temperatures diffusion into crystallites bulk is possible, as well as grain boundary migration, which already becomes apparent in the structure. As a result, the volume line intensity increases and the grain-boundary line intensity decreases.

The comparison of spectral lines relative intensities for poly- and nanocrystalline Nb demonstrates that at one and the same temperature the volume line relative intensity for the latter is much higher than for the former. It means that in case of nanocrystalline Nb, Sn atoms more readily transfer from a grain-boundary into the bulk, which may be due to high concentration of non-equilibrium vacancies in near-boundary areas. Besides, in case of nanocrystalline Nb one should not exclude the probability of grain boundary migration even at the lowest annealing temperatures, while in case of polycrystalline Nb it is evidently impossible.

Figure 19 demonstrates temperature dependences of isomer shifts and widths of ^{119m}Sn spectral lines in poly- and nanocrystalline niobium. Dashed line shows the isomer shift in Nb regular lattice determined for single-crystal niobium.

For grain-boundary lines the parameters of the Mössbauer spectra in poly- and nanocrystalline Nb are practically the same, testifying that the states of ^{119m}Sn atoms in grain boundaries differ only slightly, that is the structure of grain boundaries and the mechanism of Sn diffusion in them are similar. On the contrary, the state of ^{119m}Sn atoms in near-boundary areas of poly- and nanocrystalline Nb appreciably differs. In polycrystalline Nb the isomer shift of component 2 is higher than in the regular lattice, while for the nanostructured Nb the situation is the opposite. It means that the electron density on the Mössbauer isotope nuclei in near-boundary areas of nanocrystals is higher than in the regular lattice, and it may be stated that in these areas Sn atoms are in the compressed state, which may be due to their high defectiveness. The volume line in the nanocrystalline Nb spectra is much wider than in the polycrystalline, testifying larger range of possible states of Sn atoms in near-boundary areas of the former.

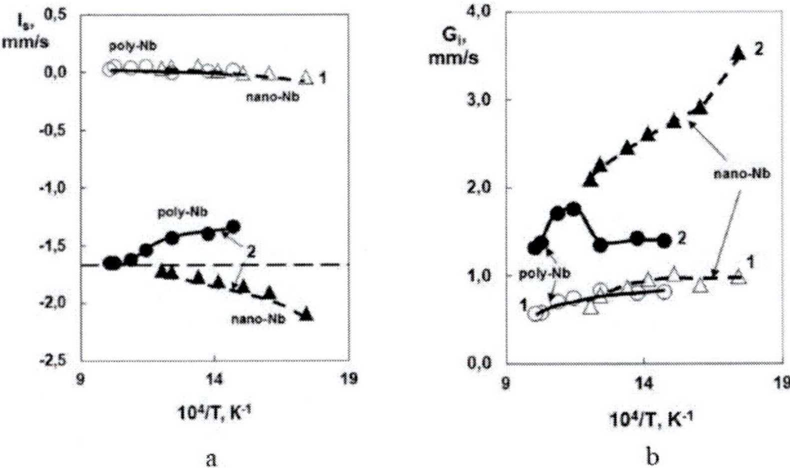

Figure 19. Temperature dependences of the isomer shifts (a) and widths (b) of components 1 and 2 of ^{119m}Sn lines for poly- and nanocrystalline Nb. The horizontal dashed line shows the isomer shift of the ^{119m}Sn line for single-crystal Nb.

With the annealing temperature growth the volume line isomer shift in the spectrum of nanocrystalline Nb increases, approaching gradually to its value in the regular lattice. It starts to grow at the lowest annealing temperatures testifying modification of the state of atoms located in near-boundary areas of nanostructured Nb. This means that the state of grain boundaries in nanostructured Nb approaches to equilibrium.

In [68] we carried out NGR studies of grain boundaries in Nb after 16 passes of ECAP. As demonstrated above, the highly non-uniform grain-subgrain structure is formed in Nb after such treatment. In the emission Mossbauer spectra of ^{119m}Sn, taken from these specimens, not two but three components were found at all the annealing temperatures studied. Judging from the parameters of these lines, we came to a conclusion, that component 1 corresponds to the atoms localized in grain boundaries, whereas the presence of two "volume" lines instead of 1 is due to the above-mentioned structural non-uniformity after ECAP and the presence of near-boundary areas of two types, namely, at equilibrium and non-equilibrium boundaries.

Thus application of the emission Mossbauer spectroscopy allowed us to reveal specific features of the so-called "non-equilibrium" grain boundaries in Nb subjected to SPD.

IV. PECULIARITIES OF STRUCTURE AND TEXTURE OF NB IN HIGH-STRENGTH NANOCOMPOSITES CU-NB

One of the most important applications of niobium is using it as a constituent of heavily-deformed Cu-Nb composites which exhibit extremely high strength along with excellent conductivity and attract great attention of the researchers as unique electro-technical material [8, 9]. Nb can serve as the second constituent of such composites due to negligible mutual solubility of copper and niobium. It is also of importance that their melting temperatures considerably differ (T_m is 1356K for copper and 2740K for Nb). At mechanical processing of ductile two-phase mixtures of Cu and Nb, containing about 20% of the latter, niobium dendrites are elongated in Cu matrix. These composites are referred to as *in situ* because both phases are deformed simultaneously, and filaments of Nb phase with very large aspect ratios are formed in Cu matrix. These alloys can be heavily deformed by cold drawing up to extremely high degrees, the thickness of the Nb filaments reducing to several nm. In a heavily-drawn condition the Nb filaments acquire a ribbon-like form and a sharp fiber texture with <110> axis.

Cu-Nb alloys containing up to 30 % Nb (volume fraction) can be cold worked by drawing or rolling with as high true strain, e, as 12 [69]. At drawing Nb dendrites transform into the ribbon-like filaments with axial texture of <110> type, which is, according to [70], due to peculiarities of <110> texture development in BCC metals, when only two of four <111> directions of sliding are oriented favorably to receive tension parallel to the filament axis. Consequently, the deformation is flat instead of the axially-symmetric flow, and the transverse sections of filaments are rectangular or elliptical, but not round. Such deformation mechanism was firstly suggested for heavily drawn polycrystalline Nb, in which it was found that every grain constrained by neighboring grains acquires the ribbon-like shape and is curled about the wire axis [71]. This mechanism is even more pronounced in two-phase alloys, in which one phase is BCC, whereas another one is FCC. The BCC filaments have to curl and bend because of compression from the surrounding FCC matrix which, on the contrary, has axially symmetric flow.

One of the direct consequences of such deformation behavior is that the effective diameter of a filament is smaller and the interface filament/matrix surface is larger than in case of axially symmetric flow. At high enough strains the strength of such composites reaches extremely high values well exceeding the rule of mixtures predictions [72, 73]. Though niobium proper is not a highly strong metal, in these composites at small enough filament sizes the presence of about 10% of Nb results in increasing of ultimate tensile strength (σ_u) of copper from ~ 345 to 1345 MPa, and 18% of Nb to 2230 MPa [70], this latter value being as high as that of the best copper whiskers.

As the rule of mixtures cannot adequately predict mechanical properties of ultra-fine multifilamentary composites, special strengthening mechanisms should be worked out for them, taking into account the presence of interfaces and matrix-filament compatibility. Two main mechanisms explaining extremely high strength of *in situ* composites are referred to as substructural and barrier models.

The model of substructural strengthening [74, 75] attributes the excess strengthening of composites to generation of additional geometrically necessary dislocations compared to similarly deformed single-phase materials for the accommodation of strain at BCC/FCC interfaces. According to the barrier mechanism [72, 73, 76], high strength results from difficulty of propagating plastic flow through FCC/BCC interfaces.

In the substructural model it is stressed that the greater is the mismatch between lattices of a composite constituents, the higher is the geometrically necessary dislocations density, the amount of which drastically increases with strain.

Some fitting parameter characterizing the mismatch degree is introduced, and it is emphasized that the greatest mismatch, and hence the highest strengthening, is attained in composites with different crystal lattices of phases, particularly, BCC-FCC. Finally, this model gives the strength dependence of Hall-Petch type, i.e. the strength depends linearly on $\lambda^{-1/2}$, where λ is an average filament spacing. An analogous conclusion can be made from the barrier model, though it is based on other fitting parameters. The main argument of the authors of the barrier mechanism against the substructural strengthening is that in real composites such high dislocation density (higher than $10^{13}/cm^2$), that is predicted by the latter, is never achieved, and, moreover, with strain growth the strength increases though the dislocation density does not grow, but ever decreases, especially in copper matrix which undergoes dynamic recovery and recrystallization [72]. That is why the barrier model is based on the inter-filament spacing, and filaments are considered as flat barriers preventing dislocation movement and hampering plastic flow between two phases.

A physical model of strengthening of such composites referred to as the "modified rule of mixtures" is suggested in [77], and it takes into account both dislocation structure and crystallographic texture of both phases. The authors of this model emphasize that to understand the high strength of *in situ* composites one must take into account the microstructure of both constituents, especially the substructure of Nb filaments, in which they found dislocation cells, dislocation pile-ups and glass-like (almost amorphous) areas practically free from dislocations.

In spite of great interest in Cu-Nb composites and quite a few publications, there are a number of debatable questions obscuring understanding of the processes responsible for their unusual behavior and, thus, preventing from their further development. There is great disagreement about dislocation densities in both constituents, especially in Nb filaments; there is a lack of information on the texture of such composites; their behavior at annealing is not well studied; potentialities of their further strengthening by doping are not well known yet, etc. That is why the authors of this chapter (with coauthors) for many years studied Cu-Nb composites fabricated in Bochvar High-Technological Institute of Inorganic Materials [78-85], and the most interesting results of these studies, together with the available data of other authors, are discussed in this section.

Let's consider the evolution of structure of Cu-Nb composites at cold drawing and intermediate annealing, beginning from specimens cold drawn from the diameter of 30 mm (after extrusion) to 2mm. The true strain, e, is in this case 5.4, and the structure is shown in Figure 20. In the transverse section (Figure 20a) it is clearly seen that Nb filaments have already acquired the pronounced ribbon-like shape, and they bend and curl around the drawing axis in agreement with previous publications and above-mentioned peculiarities of BCC metals behavior at drawing. Because of intricate shape of Nb ribbons their sizes and size scattering can hardly be evaluated, but it can be noted that the ribbons thickness is about 30-60 nm, and their width is up to several microns. According to [86], this wide size scattering is explained as follows. At low strains (up to $e = 3$) dendrite axes of the first order arrange along the drawing axis, a part of them remaining practically non-deformed because of non-uniform deformation distribution, namely, the true strain of Nb filaments is lower than that of copper matrix.

At the true strain of $e > 3$ the second- and third-order dendrite axes start to deform and elongate parallel to the drawing axis, and only at much higher strains ($e > 6$) Nb filaments get more uniform in shape and sizes.

In the copper matrix one can see nearly equiaxed grains the size of about 300 nm, characteristic of recrystallized state. Such structure may result from dynamic recrystallization of copper.

In longitudinal section the structure looks differently. In the bright-field images one can see strongly elongated parallel to the drawing axis Nb filaments with Cu layers between them. As demonstrated by the analysis of a number of dark-field images, Nb filaments considerably differ in width (which is also obvious in the transverse sections). In most electron diffraction patterns there are simultaneously reflections belonging to two reciprocal lattice planes, $(100)_{Nb}$ and $(111)_{Nb}$, and their common $<110>_{Nb}$ direction coincides with drawing direction and filament boundaries (Figure 20c).

Thus, it may be concluded, firstly, that fiber texture of Nb filaments is forming, and secondly, that every filament is subdivided into several strongly elongated grains with certain orientation relatively to each other and to plane of the flat ribbon-like filament proper. The dark-field image of one of such grains the width of 200 nm is shown in Figure 20b.

As stated in [87], in longitudinal sections one and the same filament may *seem* split into several bands parallel to the drawing axis due to their irregular shape in transverse sections. On the other hand, the authors of the cited publication also observed simultaneously two zone axes, $[001]_{Nb}$ and $[111]_{Nb}$, from different Nb grains in the electron diffraction patterns, which means that a filament can really consist of several grains with certain orientation. This interesting phenomenon is discussed in more detail below, for composites with higher strain.

A specific feature of the internal structure of Nb grains in the composite under study is the presence of very fine (of about 10-15 nm) blocks clearly seen in the dark-field images (Figure 20b). The dislocation density is high, and they are mainly concentrated in the boundaries of these fine blocks.

As the fabrication of Cu-Nb composites with ultra-fine filaments is an extremely complicated multi-stage technological process, and the material is strengthened at every pass, intermediate annealing is required to avoid its fracture. It is of great importance to know the effect of annealing on the structure and properties of the composite wire, keeping in mind that thermal stability of structure in severely deformed materials is decreased.

After the annealing at $800^{\circ}C$, 1 h, which is used in fabrication of such wires, the structure of Cu-Nb composite has an appearance demonstrated in Figure 21. It is obvious that this annealing results in marked changes of the structure, as Nb filaments are coarsened, and their shape changes from flat to more rounded. Besides, dislocation density decreases, the blocks disappear, and the filaments get more uniform in sizes, their average width being of about 250 nm.

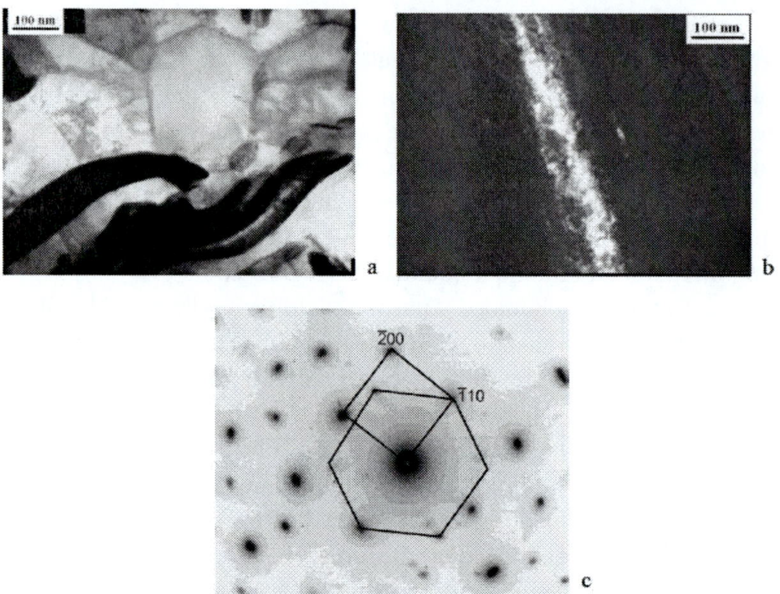

Figure 20. Transverse (a) and longitudinal (b,c) sections of Cu-Nb composite deformed to $e = 5.4$: a – light-field image; b – dark-field image in ($\overline{2}00$)$_{Nb}$ reflection; c – electron diffraction pattern, zone axes $[001]_{Nb}$ and $[111]_{Nb}$.

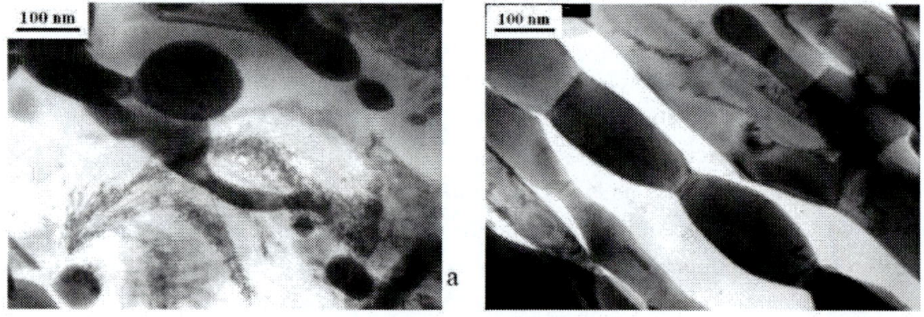

Figure 21. Transverse (a) and longitudinal (b) sections of *in situ* Cu-Nb composite deformed to $e = 5.4$ and annealed at 800°C, 1 h.

A similar effect of annealing, that is, coagulation and spheroidization of Nb filaments, accompanied with decreasing of strength, was found in a number of publications [88, 89]. Thus, for example, in [88] the evolution of Nb filaments morphology in *in situ* composites was studied in the process of manufacturing of Nb_3Sn-based superconducting wires. To obtain the superconducting Nb_3Sn compound the composites were covered with Sn and annealed, and the evolution of the shape and sizes of Nb filaments was studied. The authors of the cited publication observed modification of their morphology, namely, from flat to rounded, with waists and even ruptures, which they referred to as "sausaging".

Though the majority of the available publications mention the Nb filaments coagulation at annealing, there is certain disagreement about the temperature at which modification of filaments gets noticeable and is accompanied with changes in mechanical characteristics. For example, in [89] it is reported about the initial stages of coagulation (filament speroidization and breakage) even at the lowest annealing temperature, 250°C, 10 h. The authors found that the process begins from the thinnest filaments, then it is further developed in wider filaments at 400°C, 10 h annealing, and after the 650°C, 10 h all the filaments are modified. In [90] it is reported about possible coagulation at 350°C, 24 h annealing. Decreasing of strength observed in [91] at annealing higher than 300°C is also attributed to filament coagulation, and it is noted that this effect is more pronounced with strain growth in composites with 5 and 20 % Nb. To determine the temperature of the beginning of structural changes at annealing of *in situ* Cu-Nb wires we compared the structure of the composite cold drawn to the diameter of 0.67 mm ($e = 7.6$) and then annealed at 300 and 600°C for 1 h. As seen from Figure 22, in the composite with $e = 7.6$ Nb filaments consist of grains, elongated in the drawing direction and separated with flat high-angle boundaries. Analysis of electron diffraction patterns shows that the drawing direction and elongation axis of Nb filaments coincide with crystallographic direction of $<110>_{Nb}$, i.e., the classical axial (fiber) texture for BCC metals is developed in Nb, and it is the sharper the higher is the strain.

Analyzing a great number of electron diffraction patterns, we have found that the filament plane most often coincides with Nb lattice planes {100}, {111} and {311} (Figure 22d). Thus, in neighboring Nb grains these planes are parallel to each other and to that of the ribbon-like filament plane. In the dark-field image taken in the reflection belonging to all these planes (indicated by aperture in Figure 22d) all the filament is seen, with pronounced blocked structure, i.e. the grains consist of fine blocks separated with low-angle boundaries (Figure 22a). In the dark-field images shown in Figure 22b,c the grains with one common direction [110] are seen, for which the (111) and (100) planes coincide with the filament plane, respectively.

Thus, it can be concluded that the limited texture, characteristic of the rolling texture of BCC Nb, with the following main components, {100}<110>, {111}<110> and {311}<110> is formed in the ribbon-like Nb filaments.

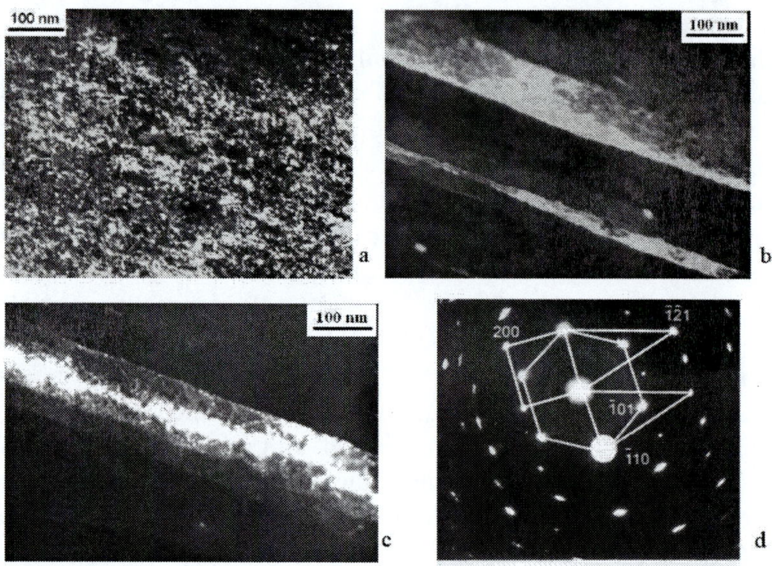

Figure 22. The structure of *in situ* Cu-Nb composite deformed to $e = 7.6$ ($\varnothing$ 0.67 mm): a – dark-field image in ($\bar{1}$10)$_{Nb}$ reflection, indicated by aperture; b – dark-field image in ($\bar{1}$01)$_{Nb}$ reflection; c - dark-field image in (200)$_{Nb}$ reflection; d – electron diffraction pattern, zone axes [111]$_{Nb}$ || [001]$_{Nb}$ || [113]$_{Nb}$.

The development of the limited texture within every filament is obviously because the BCC filaments deformation is not axial symmetric and they have the ribbon-like shape. The fact that the filaments consist of several grains separated by high-angle boundaries is probably the result of dislocation rearrangement in the filaments which had already acquired the ribbon shape. A similar effect was observed in [76] in heavily rolled Cu-Nb composites, where it was observed that at the true strain of 6.0 dislocations are distributed chaotically, but with e growth up to 6.9 they rearrange into low-angle boundaries parallel to <110>$_{Nb}$. In [92] Nb filaments chemically extracted from Cu-Nb composite were studied, and it was found that at e higher than 7 the filaments have boundaries almost parallel to <110>$_{Nb}$, which separate areas with relative misorientation of 2-35°.

It may be also suggested that one of the reasons of the limited texture formation in Nb filaments is the influence of the surrounding FCC copper matrix which adjusts to axial-symmetric flow.

After the annealing at 300°C, 1 h there are no marked changes in the structure compared to the deformed state, and it cannot be stated that this annealing results in coagulation of Nb filaments. The fine-blocked structure is retained in the filaments, though dislocation density is somewhat lower. The Nb filaments are more uniform in sizes after the annealing, their thickness being about 90 nm with copper layers of 120 nm width between them, though in some areas the filament size scattering is quite pronounced.

The annealing temperature increasing up to 600°C results in noticeable changes of structure (Figure 23). The blocked structure of the Nb filaments disappears, and the dislocation density decreases considerably. Besides, the shape of Nb filaments changes, constrictions and convexities appear, indicating intensive coagulation at this temperature. The filament size distribution gets more uniform, their average width being 120 nm.

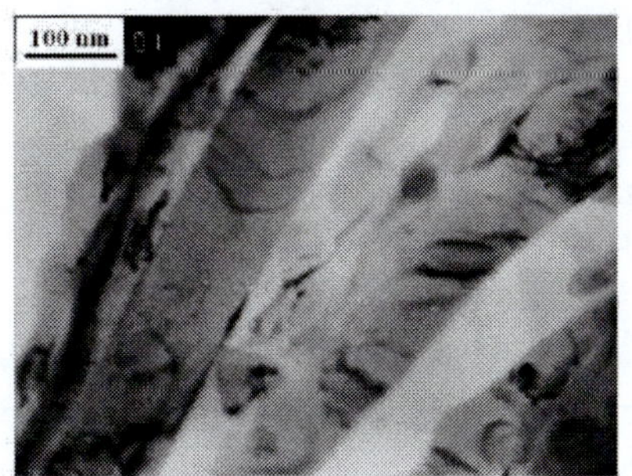
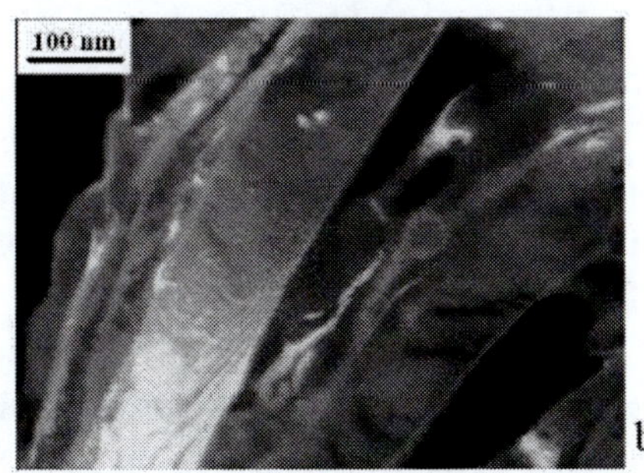

Figure 23. The structure of Cu-Nb composite deformed to $e = 7.6$ and annealed at 600°C, 1 h: a –bright-field image; b – dark-field image in $(110)_{Nb}$ reflection.

Though it is quite obvious from the electron-microscopic images that the dislocation density in Nb filaments markedly decreases at annealing, especially at higher temperature of the latter, it was of interest to evaluate this very important characteristic quantitatively and to determine its influence on the strength. That is why we determined the dislocation density by the X-ray method and measured the microhardness as an indirect characteristic of strength.

The problem of dislocation density evaluation is of great importance for revealing the reasons of unusual properties of heavily-deformed composite materials. The main disagreement of the above-mentioned models of barrier and substructural strengthening concerns the dislocation density. The main argument of the authors of the barrier mechanism against the substructural is that the latter implies its continuous growing by a factor of two in the true strain range from 3.1 to 11.9, whereas according to the experimental data it remains almost constant [93].

Discrepancies of different authors in dislocation density values are very high (several orders of magnitude). Thus, for example, in [70] high strength of *in situ* composites is attributed to extremely high dislocation density of 10^{12}-10^{13} cm^{-2}. On the contrary, considerably lower values of about 10^{10} cm^{-2} were obtained by electron microscopy of Cu-Nb composites in [72]. The problems of dislocation density evaluation in these materials by TEM are considered in detail in [94]. They are caused, first of all, by high dispersity of heavily-deformed composites, the structure of which possesses elements the sizes of less than 100 nm, which requires very high magnifications. Secondly, the composites considered are two-phase, which makes difficult to obtain good enough thin foils, as Cu and Nb behavior at etching appreciably differs, and it is not always possible to make both constituents thin enough simultaneously.

Based on the detailed analysis of dislocation structure of Cu matrix, it was concluded in [94] that at all deformations studied its dislocation density is 10^{10} - 10^{11} cm^{-2}, and it does not change systematically with strain growth. Besides, dislocations are distributed non-uniformly, some grains being practically dislocation-free, whereas in dislocation cells the dislocation density is an order of magnitude higher.

Unambiguous conclusion on evolution of microstructure of *in situ* composites at deformation and annealing can be done based on combination of the data of various research methods, particularly, transmission and scanning electron microscopy, X-ray analysis and conductivity measurement.

X-ray analysis of Cu-Nb specimens after deformation and annealing was done on DRON-3M diffractometer in CuKα radiation with Ni filter. The lines (200) for Nb and (111) for Cu were registered. The same lines were registered for standard specimens for which the well-annealed slices of pure copper and niobium were used. The dislocation density was determined as:

$$\rho = (\beta^2 / 2b^2) ctg^2 \theta_0 , \tag{8}$$

where β is the integral physical line width in radians, b is Burgers vector, and θ_0 is the reflection angle of an ideal crystal.

The data obtained together with microhardness are given in Table 4.

Table 4. Dislocation density (ρ) and microhardness (H) of Cu-Nb composites after deformation and annealing

№	$\varnothing$, mm	Обработка	ρ_{Nb}	ρ_{Cu}	H, МПа
1	0.67	drawing to $e = 7.6$	$5.0 \cdot 10^{11}$ cm^{-2}	$6.0 \cdot 10^{10}$ cm^{-2}	2970
2	0.67	drawing to $e = 7.6$ + annealing 300°C/1h	$3.5 \cdot 10^{11}$ cm^{-2}	$5.5 \cdot 10^{10}$ cm^{-2}	2250
3	0.67	drawing to $e = 7.6$ + annealing 600°C/1h	$1.3 \cdot 10^{11}$ cm^{-2}	$5.5 \cdot 10^{10}$ cm^{-2}	2100

As seen from the table, the dislocation density in Nb filaments cold-drawn to $e = 7.6$ is $5.0 \cdot 10^{11}$ cm^{-2}, and in copper matrix it is $6.0 \cdot 10^{10}$ cm^{-2}. These results are in agreement with the data presented in [72, 73, 76, 94], and they disagree with the authors who indicate much higher dislocation densities (e.g. [70]).

Annealing at 300°C, 1 h causes the dislocation density decreasing both in Nb and Cu. At higher temperature annealing (600°C, 1 h) the dislocation density in Nb decreases further, but in Cu it remains practically unchanged. As shown by the above described electron-microscopic investigations, at 600°C coagulation of Nb filaments is observed, and in Nb/Cu interfaces the recovery proceeds, which is accompanied by the decreasing of dislocation density due to their annihilation. At lower temperature, 300°C, these processes only start and are not as pronounced.

Annealing at 300 and 600°C results in marked drop of microhardness, which correlates with decreasing dislocation density. Thus, the dislocation density decreasing both at Nb/Cu interfaces and in Nb filaments bulk, coagulation of Nb filaments and disappearance of blocked structure in them, promote the composite softening, which is reflected in lower microhardness.

The obtained data on the effect of annealing on the structure and strength should be taken into account when the regimes of intermediate annealing are chosen for manufacturing of high-strength *in situ* Cu-Nb composite wires, keeping in mind that even at as low as 300°C certain changes of structural and mechanical characteristics are possible in heavily-deformed material.

Let's consider in more detail the evolution of structure and texture of Cu-Nb composites with strain growth by routes A and B differing in intermediate annealing.

Route A: Drawing to $\varnothing$ 10 mm + annealing at 700°C, 1 h + drawing to final diameter (1 or 0.3 mm) (the total true strain, e, comprises 6.80 and 9.21 respectively).

Route B: Drawing to $\varnothing$ 10 mm + annealing at 700°C, 1 h + drawing to 3 mm + annealing at 700°C, 1 h + drawing to $\varnothing$ 1 mm + annealing at 550°C, 1 h + drawing to final diameter (0.8 or 0.3 mm) (the total true strain, e, comprises 7.27 and 9.21 respectively).

The texture and structure of these specimens were studies by the X-ray analysis in DRON-3M diffractometer with CuKα radiation. The texture was studied by construction and analysis of inverse pole figures for Nb filaments and Cu matrix. The pole density was calculated according to:

$$P_{hkl} = \frac{\left(I_{T,HKL} / I_{0,HKL}\right)}{\sum\limits_{n}\left(Z_{hkl}I_{T,HKL} / I_{0,HKL}\right)} \sum\limits_{n} Z_{hkl} \tag{9}$$

where hkl are the indexes of a plane, HKL are the interference indexes, $I_{T,HKL}$ is the interference peak intensity for a textured specimen (index T denotes "textured") and $I_{0,HKL}$ is the interference peak intensity for a non-textured standard (index 0 denotes "non-textured"), and Z_{hkl} is the reiteration factor of a crystallographic plane with hkl indexes.

The intensities of the reflections from 5 crystallographic Cu planes, (100), (110), (111), (311), (331), and 6 Nb planes, (100), (110), (111), (211), (310), (321), have been determined in the construction of the inverse pole figures. It was not possible to increase the number of the analyzed reflections as the investigated specimens were the two-phase ones and, besides, because all the lines were broad due to a high deformation degree, and it was difficult to attain good separation of different reflections.

Diffraction patterns obtained from the cross-sections of all specimens demonstrate only reflections from $\{110\}_{Nb}$ planes for Nb. Other reflections were not observed at all, testifying the development of the perfect single-component fiber texture in the Nb filaments. The $<110>_{Nb}$ axis coincides with the drawing direction, in agreement with previous publications [77, 86] and the above described electron-microscopic data.

The intermediate annealing could not noticeably affect the texture, as the annealing temperature was appreciably lower than the temperature of recrystallization of niobium weakening the texture. The dynamic recrystallization in niobium has not been discovered as well even at the highest investigated drawing ratios.

The texture characteristics of copper matrix are intricately affected by changes in deformation degree and introduction of intermediate annealing for three possible reasons. Firstly, the two-component fiber structure can develop in Cu matrix, with $<111>_{Cu}$ and $<100>_{Cu}$ axes [92]. Secondly, at the intermediate annealing used the recrystallization of the Cu matrix is quite possible, as for copper, contrary to niobium, the annealing temperature was high enough. Thirdly, at high deformation degrees dynamic recrystallization of the Cu matrix is possible as well [73]. Thus, the effect of the drawing ratios and intermediate annealing on the Cu matrix texture should be analyzed in view of their influence on the microstructure.

Figure 24 demonstrates inverse pole figures presented as standard stereographic triangles for the specimens studied, and the pole densities for the two possible fiber texture components of the copper matrix, $<111>_{Cu}$ and $<100>_{Cu}$. As seen from the figure, the $<111>_{Cu}$ fiber texture axis dominates in the Cu matrix of the cold drawn (with 1 intermediate annealing, route A) specimens, though there is also a weak $<100>_{Cu}$ orientation (= 1,1). A very high degree of texture is attained at $e = 6.8$, but at further growth of strain up to 9.21 the texture weakens (P_{111} decreases from 7.4 to 3.7, and P_{100} drops down to 0.6, i.e. it practically vanishes). It is known that in cold drawn pure copper fiber texture with two axes, $<111>_{Cu}$ and $<100>_{Cu}$, develops, and the contribution of both of them is commensurable. As mentioned above, these two texture axes were observed in the Cu matrix of the heavily cold drawn Cu-Nb composites by many authors. However, a number of publications report that only one axis of fiber texture is characteristic of the Cu matrix of *in situ* composites, namely, $<111>_{Cu}$ [77]. This contradiction may be explained as follows. At intermediate strains by cold drawing both abovementioned components of texture do develop, with their approximately equal contribution [95], but the $<100>_{Cu}$ component remains at a low level with the growth of strain, whereas the contribution of $<111>_{Cu}$ axis increases, and, as a result, in heavily-drawn composites some authors revealed only the latter one.

The weakening of the texture with the growth of strain, observed in the present study, is, in the first sight, a somewhat unexpected result, and to explain it we should analyze the effect of the deformation degree on the microstructure parameters of the copper matrix. According to the X-ray data obtained, with the growth of e from 6.8 to 9.21 the size of the coherent scattering areas in Cu somewhat decreases, but is still on quite a high level of about 100 nm, whereas dislocation density practically does not change. The latter testifies dynamic recovery and recrystallization of the copper matrix at high deformation degrees, in agreement with [72, 73, 92], where it is stated that at high strains ($e > \sim 7$) an equilibrium between the nucleation and annihilation of dislocations is attained, and intensive development of the dynamic recrystallization is observed. According to [72], the cell structure is forming at intermediate strains of about 5, and at further deformation it transforms into the subgrain structure as a result of the dynamic recrystallization. The mechanism of the latter consists in the transformation of cell walls into subgrain and grain high-angle boundaries, i.e. the dynamic recrystallization of the Cu matrix promotes subgrain disorientation. Thus, it may be suggested that it is the dynamic recrystallization that is responsible for the observed weakening of the texture at high deformation degrees.

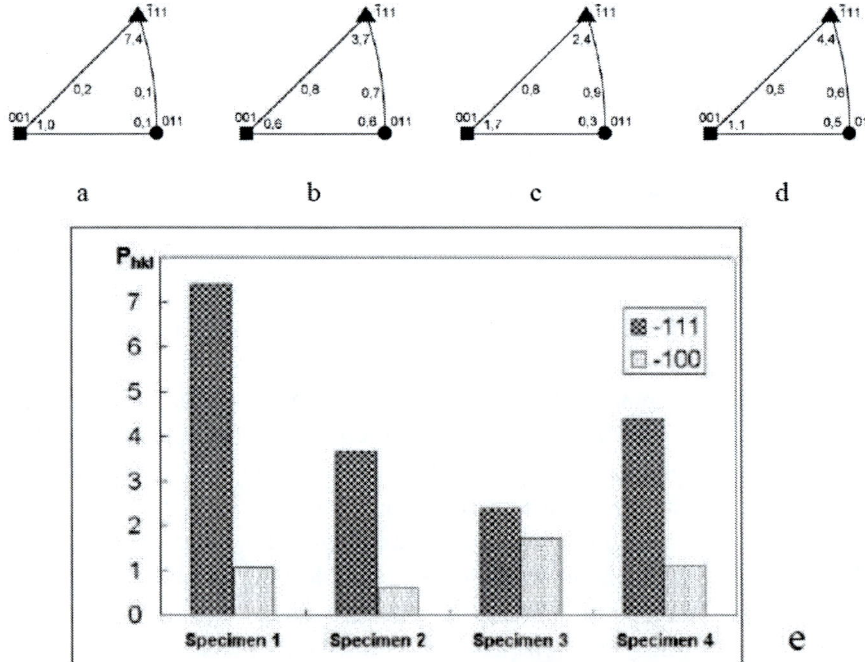

Figure 24. Inverse pole figures after deformation by Route A (a - e = 6.8, sample 1; b - e = 9.21, sample 2) and Route B (c - e = 7.25, sample 3; d - e = 9.21, sample 4) and pole densities of $<111>_{Cu}$ and $<100>_{Cu}$ axes in the drawing direction (e).

In the specimens processed with additional intermediate annealing (Route B) there are the same two components of the fiber texture in the Cu matrix,

$<111>_{Cu}$ and $<100>_{Cu}$, their contribution being commensurable at $e = 7.25$ and the texture being not sharp ($P_{111} = 2.4$, $P_{100} = 1.7$) (specimen 3). With the growth of strain up to 9.21, the P_{111} increases up to 4.9, whereas the P_{100} decreases to 1.1, i.e. the $<100>_{Cu}$ component practically vanishes (specimen 4).

Processing by route B results in the substantially lower dislocation density of the Cu matrix compared to route A, testifying that in the case of deformation with intermediate annealing the equilibrium between the nucleation and annihilation of dislocations is not attained, and the development of the dynamic recrystallization is not possible. This may be explained by the fact that the annealing temperatures are sufficient for the development of the recovery and recrystallization in the Cu matrix, and it is not the total strain but the deformation after the last annealing that affects the formation of the microstructure and texture in this case. In specimen 3 processed by route B to $\varnothing$ 0.8 mm the total strain is high ($\eta = 7.25$), but the strain after the last annealing at $\varnothing$ 1 mm is only 0.44, and it's not surprising that the dislocation density of the Cu matrix in this case is well below that of the specimens processed by route A. By the same reason, the contributions of both texture components are commensurable, and the texture is not very perfect. When the strain after the last annealing increases up to 2.41 (the total strain being 9.21), the $<111>_{Cu}$ component gets stronger, whereas the $<100>_{Cu}$ axis weakens and the dislocation density increases. As a result, the parameters of the Cu matrix texture and microstructure in that case approach to the characteristics of the cold drawn (by route A) wire, though don't completely reach them.

Thus, the texture of *in situ* Cu-Nb composite wire has been studied by the X-ray analysis, along with the electron microscopy, after various regimes of deformation and annealing. The sharp single-component fiber texture with $<110>_{Nb}$ axis has been found to develop in Nb filaments, in agreement with previous publications. The introduction of the additional intermediate annealing does not affect the Nb filaments texture, because the annealing temperature is well above the temperature of the recrystallization of niobium, and the dynamic recrystallization is not observed in the Nb filaments as well, even at the highest strain studied ($e = 9.21$). Two components of the fiber texture, $<111>_{Cu}$ and $<100>_{Cu}$, have been discovered in the Cu matrix of the composites investigated. These components were found to change intricately with deformation and annealing. After the cold drawing with e up to 6.8 the $<111>_{Cu}$ component is strong and the $<100>_{Cu}$ is weak. Both of them are weakened with the growth of deformation up to 9.21 due to the dynamic recovery and recrystallization. The introduction of the intermediate annealing appreciably affects the texture of the Cu matrix. At low deformation degree after the last annealing both abovementioned components are observed, and their contributions are approximately equal and the degree of the texture is not very high. With the growth of the deformation after the last annealing the $<111>_{Cu}$ component noticeably grows, while the $<100>_{Cu}$ axis practically vanishes, and it may be concluded, that the annealing results in such a substantial decrease of the dislocation density, that at further deformation it does not attain the values at which the dynamic recovery and recrystallization would be possible.

CONCLUSION

The problems of nanostructuring niobium by various techniques of severe plastic deformation are considered in this chapter. The evolution of Nb structure at ECAP, HPT and combination of these methods has been studied along with the effect of temperature on the structure refinement. It is demonstrated that when Nb is deformed by ECAP, its structure refines to submicrocrystalline sizes and is non-uniform, with grain boundaries of two types, equilibrium and non-equilibrium. The latter is evidenced by the emission Mossbauer spectra, which contain three components instead of two lines usually observed in polycrystalline metals.

Deformation of Nb by HPT at room temperature enables to obtain the structure borderline between nano- and submicrocrystalline, with grain sizes of about 120 nm. Combination of ECAP and HPT does not result in further grain refinement, and it may be concluded that the initial state (single crystalline, polycrystalline or preliminary ECAP-ed) does not affect the final grain sizes after HPT when the saturation of fragmentation is reached.

To suppress the relaxation processes and achieve further grain refinement the HPT was carried out at cryogenic temperature (in liquid nitrogen), and the uniform nanocrystalline structure was obtained, with record-breaking for niobium microhardness of 4800 MPa. This structure is stable at room temperature and does not degrade for several months ageing.

The thermal stability of Nb structure, obtained by HPT at room temperature, decreases with the growth of strain. After HPT by 5 revolutions the structure is stable up to 600°C, and after 10 revolutions the grain growth starts at 500°C annealing. The nanocrystalline Nb obtained by HPT at cryogenic temperature, possesses even lower thermal stability.

Recrystallization processes start in it at as low as 200-300°C, at 400°C marked grain growth is observed, and the structure gets submicrocrystalline, and at 500°C complete recrystallization occurs.

By the method of the emission Mössbauer spectroscopy the specific features of grain boundaries in Nb nanostructured by HPT compared to that in ordinary polycrystalline state have been revealed. These features are caused not by small grain sizes, but by high defectiveness of boundaries after SPD. According to the Mössbauer spectroscopy data, the near-boundary areas in nanocrystalline Nb are enriched in non-equilibrium vacancies and are characterized with larger set of possible states in which the diffusing atoms can locate.

As Nb is one of the two main constituents of high-strength *in situ* Cu-Nb nanocomposites, which are studied and used as a very important electro-technical material, the results of studies of their evolution at various regimes of deformation and annealing are presented in this chapter as well. The multiple cold drawing of these materials to as high true strain as 11 is possible without their fracture, and the structure of Nb filaments and Cu layers between them refines to nanometer sizes. The Nb filaments acquire the ribbon-like shape which is caused by specific behavior of BCC metals at drawing, as well as by the effect of the surrounding Cu matrix. The sharp fiber texture of $<110>_{Nb}$ type is developed in the composite wires. In Cu matrix the $<111>_{Cu}$ directions are also parallel to the drawing axis, but between these two directions, $<110>_{Nb}$ and $<111>_{Cu}$ there is always an angle of about 2-5°, which indicates the partly coherent connection between the filaments and matrix. Along with the sharp fiber texture, in Nb filaments the limited texture with $\{111\}<110>$, $\{100\}<110>$ and $\{311\}<110>$ components is formed, characteristic of the cold-rolled Nb, which is due to their ribbon-like shape. Intermediate annealing at cold drawing of Cu-Nb composites result in coagulation of Nb filaments, which starts at as low as 300°C, and at 600°C, 1 h is well pronounced and causes the loss of strength.

The comparison of Nb behavior in the free state and in composites, at different deformation modes, including severe plastic deformation, revealing of specific features of the state of grain boundaries, determining of mechanisms of grain-boundary diffusion can contribute to deeper understanding of the processes which provide the basis for creation of various advanced materials based on Nb and allow to find ways of their further optimization.

REFERENCES

[1] Gleiter, H. *MRS Bull.* 2009, vol.34, 456-463.

[2] Valiev, R.Z.; Aleksandrov, I.V. Bulk nanostructured metallic materials: manufacturing, structure and properties. Moscow: *"ACADEMKNIGA"* (in Russian), 2007, 398 p.

[3] Gusev, A. I.; Rempel, A.A. Nanocrystalline Materials. Cambridge: *Cambridge International Science Publishing,* 2004. 351 p.

[4] Glezer, A.M.; Permyakova, I.E. Nanomaterials quenched from melt. Moscow: *FIZMATLIT* (in Russian), 2012, 360 p.

[5] Würschum, R.; Herth, S; Brossmann, U. Proc. Conf. «Nanomaterials by Severe Plastic Deformation – NANOSPD2» Vienna, Austria, 2002. 755-766.

[6] Baro, M.D; Kolobov, Yu.R.; Ovidko, I.A.; Schaefer, H.E.; Straumal, B.B.; Valiev R.Z., Alexandrov I.V., Ivanov M., Reimann K., Reizis A.B., Surinash, S.; Zhilyaev, A.P. *Rev. Adv. Mater. Sci.* 2001, vol. 2, 1-43.

[7] Valiev, R.Z.; Estrin, Y.; Horita, Z.; Langton, T.G.; Zehetbauer, M.J.; Zho, Y. J. *Miner. Met. Mater. Soc.* (JOM) 2006, vol. 58(4), 33-39.

[8] Pantsyrnyi, V.I. *IEEE Trans. Appl. Supercond.* 2002, vol. 12, 1189-1194.

[9] Shikov, A.; Pantsyrnyi, V.; Vorobieva, A.; Khlebova, N.; Silaev, A. *Physica C* 2001, vol. 354(1-4), 410-414.

[10] Valiev, R.Z.; Krasilnikov, N.A.; Tsenev, N.K. *Mater. Sci. Eng. A* 1991, vol. 137, 35-43.

[11] Valiev, R.Z.; Korznikov, A.V.; Mulyukov, R.R. *Mater. Sci. Eng. A* 1993, vol. 168, 141-148.

[12] Segal, V.M. *Mater. Sci. Eng. A* 1999, vol. 271, 322-333.

[13] Ferrase, S.; Segal, V.M.; Hartwig, H.T. *Metal. Mater. Trans. A* 1997, vol. 28, 1047-1055.

[14] Mishin, O.V.; Gertsman, V.Y.; Valiev, R.Z. *Scripta Mater.* 1996, vol. 35, 873-880.

[15] Pippan, R., Scheriau, S.; Hohenwarter, A.; Hafok, M. *Mater. Sci. For.* 2008, vol. 584-586, 16-21.

[16] Horita, Z.; Furukava, M.; Nemoto, M.; Langton, T.G. *Mater. Sci. Tech.* 2000, vol. 16, 1239-1243.

[17] Dobatkin, S.V.; Kopylov, V.I.; Pippan, R.; Vasil'eva, O.V. *Mater. Sci. For.* 2004, vol. 467-470, 1277-1282.

[18]   Iwahashi, Y.; Wang, J.; Horita, Z.; Nemoto, M.; Langton, T.G. *Scripta Mater.* 1996, vol. 35(2), 143-146.

[19]   Sevillano, J.G. *Proc. 25th Risoe Int. Symp. Mater. Sci.,* Denmark, 2004, 1-11.

[20]   Popova, E.N.; Popov, V.V.; Romanov, E.P.; Pilyugin, V.P. *Phys. Met. Metallogr.* 2006, vol. 101(1), 52-57.

[21]   Popova, E.N.; Popov, V.V.; Romanov, E.P.; Pilyugin, V.P. *Phys. Met. Metallogr.* 2007, vol. 103, 407-413.

[22]   Degtyarev, M.V.; Chashchukhina, T.I.; Voronova, L.M.; etc. *Acta Mater.* 2007, vol. 55 (18), 6039-6050.

[23]   Pippan, R.; Scheriau, S., Taylor, A.; et al. *Annu. Rev. Mater. Res.* 2010, vol. 40, 319-343.

[24]   Bachmaier, A.; Hafok, M., Pippan, R. *Mater. Trans.* 2010, vol. 51(1), 8-13.

[25]   Hebesberger, T.; Stuwe, H.P.; Vorhauer, A.; Wetscher, F.; Pippan, R. *Acta Mater.* 2005, vol. 53, 393-402.

[26]   Schafler, E.; Pippan, R. *Mater. Sci. Eng. A* 2004, vol. 387-389, 799-804.

[27]   Pippan, R.; Wetscher, F.; Hafok, M., Vorhauer, A.; Sabirov, I. *Adv. Eng. Mater.* 2006, vol. 8, 1046-1056.

[28]   Lian, J.; Valiev, R.Z.; Baudelet, B. *Acta Metall. Mater.* 1995, vol. 43 (11), 4165-4170.

[29]   Popova, E.N.; Romanov, E.P.; Sudareva, S.V. *Phys. Met. Metallogr.* 2003, vol. 96, 146-159.

[30]   Mulyukov, K.Y.; Korznikova, G.F.; Abdulov, R.Z.; Valiev, R.Z. *J. Magn. Magn. Mater.* 1990, vol. 89, 207-213.

[31]   Vorhauer, A.; Knabl, W.; Pippan, R. Proc. conf. "Nanomaterials by severe plastic deformation-NANOSPD2". Dec. 2002. Vienna, Austria. P. 648-653.

[32]   Wadsack, R.; Pippan, R.; Schedler, B. *Fus. Eng. Design* 2003, vol. 66-68, 265-269.

[33]   Popov, V.V.; Popova, E.N.; Stolbovskiy, A.V. *Mater. Sci. Eng. A,* 2012, vol. 539, p. 22-29.

[34]   Alkorta, J.; Martinez-Esnaola, J.M.; Sevillano, J.G. *Acta Mater.* 2006, vol. 54, 3445-3452.

[35]   Alkorta, J.; Luis Perez, C.J.; Popova, E.N.; Hafok, M.; Pippan, R.; Sevillano, J.G. *Mater. Sci. Forum* 2008, vol. 584-586, 215-220.

[36]   Spitzig, W.A.; Trybus, C.L.; Laabs, F.C. *Mater. Sci. Eng.* A 1991, vol. 145(2), 179-187.

[37]   Valiev, R.Z.; Ivanisenko, Y.V.; Rauch E.F.; et. al. *Acta Mater.* 1996, vol. 44(12), 4705-4712.

[38]   Huang, Y.; Langdon T.G. *Mat. Sci. Eng.* A, 2003, vol. 358(1-2), 114-121.

[39]   Schafler, E.; Pippan R. *Mater. Sci. Eng.* A, vol. 387-389(1-2) Spec. Iss. 799-804.

[40]   Glezer, A.M; Metlov, L.S. *Phys. Sol. State* 2010, vol. 52(6), 1162-1169.

[41]   Popov, V.V.; Popova, E.N.; Stolbovskiy, A.V.; Pilyugin, V.P. *Mater. Sci. Eng.* A 2011, vol. 528, 1491-1496.

[42]   Popov, V.V.; Popova, E.N.; Stolbovskiy, A.V.; Pilugin, V.P. *Def. Diff. For.* 2010, vol. 297-301, 1312-1321.

[43]   Pilyugin, V.P.; Voronova, L.M.; Degtyarev, M.V.; et al. *Phys. Met Metallogr.,* 2010, vol. 110(6), 564-573.

[44]   Pilyugin, V.P.; Gapontseva, T.M.; Chashuhina, T.I.; et al. *Phys. Met Metallogr.,* 2008, vol. 105(4), 409-419.

[45]   Sutton, A.P., Balluffi, R.W. Interfaces in Crystalline Materials. *Oxford University Press,* 1995. 856 p.

[46]   Popov, V.V. Structure and properties of grain boundaries. Chapter 3 in Current Trends in Chemical Engineering Editor: J.M.P.Q. Delgado. *Studium Press LLC.* 2010. P. 49-103.

[47]   Pumphrey, P.H.; Gleiter, H. *Phil. Mag.,* vol. 32(4), 881-885.

[48]   Valiev, R.Z.; Kaibyshev, O.A.; Khananov, S.K. *Phys. Stat. Sol.* 1979, vol. 52(2), 447-453.

[49]   Valiev, R.Z.; Gertsman, V.Y.; Kaibyshev, O.A. *Phys. Stat. Sol.* 1986, vol. 97(1), 11–56.

[50]   Nazarov, A.A.; Romanov, A.E.; Valiev, R.Z. *Acta Metall. Mater.* 1993, vol. 41(4), 1033-40.

[51]   Valiev, R.Z.; Gertsman, V.Y.; Kaibyshev O.A.; et al. *Phys. Stat. Sol. A.* 1983, vol. 77(1), 97–105.

[52]   Valiev, R.Z.; Musalimov, R.Sh. *Phys. Met. Metalloghr,* 1994, vol. 78(6), 114-121.

[53]   Popov, V.V. *Phys. Met. Metallogr.* 2013, 113(13), 1257-1289.

[54]   Popov, V.V. *Def. Diff. For.* 2006, vol. 258-260, 497-508.

[55]   Popov, V.V. *Def. Diff. For.* 2009, vol. 289–292, 633–640.

[56]   Kaigorodov, V.N.; Klotsman, S.M. Pisma JETF (*Lett. J. Exp. Theor. Phys.*) 1978, vol. 28(6), 386-388. [in Russian].

[57]   Kaigorodov, V.N.; Klotsman, S.M. *Phys. Rev.* B, 1994, vol. 49(14), 9374-9399.

[58]   Popov, V.V.; Popova, E.N.; Sergeev, A.V.; Stolbovsky, A.V.; Kazihanov, V.U.; Valiev, R.Z. Diff. Def. Data Pt. A, *Def. Diff. For.* 2008, vol. 283-286, 629-638.

[59]   Popov, V.V.; Sergeev, A.V.; Timofeev, A.N.; Kovalenko, E.V.; Grabovetskaya, G.P.; Mishin, I.P. *Phys. Met. Metallogr.* 2010, vol. 109(5), 556-562.

[60]   Popov, V.V.; Sergeev, A.V.; Grabovetskaya, G.P.; Mishin, I.P. *Diff. Def. Data Pt. A, Def. Diff. For.* 2012, vol. 326-328, 674-681.

[61]   Popov, V.V. *Phys. Met. Metallogr.* 2012, vol. 113(9), 883-887.

[62]   Popov, V.V.; Kaigorodov, V.N.; Popova, E.N.; Stolbovsky, A.V. *Def. Diff For.* 2007, vol. 263, 69-74.

[63]   Popov, V.V.; Kaigorodov, V.N.; Popova, E.N.; Stolbovsky, A.V. *Bull. RAS, Physics,* 2007, vol. 71(9), 1244-1248.

[64]   Schnetweiss, O.; Cermak, J., Turek, I.; Lejcek, P. *Hyperfine Interact.,* 2000, vol. 126, 215–218.

[65]   Popov V.V. *Phys. Met. Metallogr.* 2006, vol. 102(5), 453-461.

[66]   Popov, V.V. *Sol. State Phenom.* 2008, vol. 138, 133-144.

[67]   Fisher, J.C. J. *Appl. Phys.,* 1951, vol. 22, 74-80.

[68]   Stolbovsky, A.V.; Popova, E.N. *Bull. RAS, Physics,* 2010, vol. 74(3), 358-362.

[69]   Verhoeven, J.D.; Chumbley, L.S.; Laabs, F.C.; Spitzig, W.A. *Acta Metall. Mater.,* 1991, vol. 39(11), 2825-2834.

[70]   Bevk, J.; Harbison, J.P.; Bell, J.L. *J. Appl. Phys.,* 1978, vol. 49(12), 6031-6038.

[71]   Hosford, W.F. *Trans. Metal. Soc. AIME,* 1964, vol. 230, 12-15.

[72]   Spitzig, W.A.; Pelton, A.R.; Laabs, F.C. *Acta Metall.,* 1987, vol. 35(10), 2427-2442.

[73]   Spitzig, W.A. *Acta Metall. Mater.,* 1991, vol. 39(6), 1085-1090.

[74]   Funkenbusch, P.D.; Courtney, T.H. *Acta Met.,* 1985, vol. 33(5), 913-922.

[75]   Funkenbusch, P.D.; Courtney, T.H. *Scripta Metall.,* 1989, vol. 23, 1719-1724.

[76]   Trybus, C.L.; Spitzig, W.A. *Acta Metall.,* 1989, vol. 37(7), 1971-1981.

[77]   Hangen, U.; Raabe, D. *Acta Metall. Mater.,* 1995, vol. 43(11), 4075-4082.

[78]   Popova, E.N.; Rodionova, L.A.; Popov, V.V. et al. *Phys. Met. Metallogr.,* 1997, vol. 84(5), 114-130.

[79]   Popova, E.N.; Sudareva S.V.; Popov, V.V. et al. *Phys. Met. Metallogr.,* 2000, vol. 90(2), 199-208.

[80] Popova, E.N.; Popov, V.V.; Rodionova, L.A.; Sudareva, S.V.; Romanov, E.P.; et al. *Textures and Microstructures,* 2000, vol. 34, 263-277.

[81] Popova, E.N.; Popov, V.V.; Romanov, E.P.; Sudareva, S.V.; et al. *Scripta Mater.,* 2002, vol. 46, 193-198.

[82] Popova, E.N.; Popov, V.V.; Romanov, E.P.; et al. *Phys. Met. Metallogr.,* 2002, vol. 94(1), 73-81.

[83] Popova, E.N.; Popov, V.V.; Romanov, E.P.; et al. *Scripta Mater.,* 2004, vol. 51, 727-731.

[84] Popova, E.N.; Popov, V.V.; Rodionova, L.A.; Romanov, E.P.; Sudareva, S.V.; et al. *Def. Fracture Mater.* [in Russian], 2005, No. 2, 31-35.

[85] Popova, E.N.; Popov, V.V.; Romanov, E.P.; et al. *Def. Diff. For.* 2006, vol. 258-260, 299-304.

[86] Raabe, D.; Heringhaus, F.; Hangen, U.; Gottstein, G. *Z. Metallk.,* 1995, Bd. 86(6), 405-415.

[87] Dupouy, F.; Snoeck, E.; Casanove, M.J.; et al. *Scripta Mater.,* 1996, vol. 34(7), 1067-1073.

[88] Ochiai, S.; Osamura, K.; Kitai, T.; Yamada, Y. *J. Mater. Sci.,* 1990, vol. 25(8), 85-87.

[89] Hardwick, D.A.; Rhodes, C.G.; Fritzemeier, L.G. *Met. Trans.* 1993, vol. 24A(1), 27-34.

[90] Verhoeven, J.D.; Downing, H.L.; Chumbley, L.S.; Gibson, E.D. *J. Appl. Phys.,* 1989, vol. 65(3), 1293-1301.

[91] Spitzig, W.A.; Downing, H.L.; Laabs, F.S.; et al. *Met. Trans.* 1993, vol. 24A(1), 7-14.

[92] Pelton, A.R.; Laabs, F.C.; Spitzig, W.A.; Chen, C.C. *Ultramicroscopy,* 1987, vol. 22 (1-4), 251-265.

[93] Spitzig, W.A.; Verhoeven, J.D.; Trybus, C.L.; Chumbley, L.S. *Scripta Metall. et Mater.,* 1990, vol. 24(6), 1171-1174.

[94] Trybus, C.L.; Chumbley, L.S.; Spitzig, W.A.; Verhoeven, J.D. *Ultramicroscopy,* 1989, vol. 30, 315-320.

[95] Kalu, P.T.; Brandao, L.; Ortiz, F. et. al. *Scripta Mater.,* 1998, vol. 38(12), 1755-1761.

Reviewed by Professor V.V. Kondratiev, Institute of Metal Physics Ural Branch of Russian Academy of Sciences.

Referee's comments:

The chapter contains the authors' original research of Nb, its nanostructuring and specific features in various states; and the results of other authors dealing with similar problems are reviewed as well. The chapter is sound, original, and of interest. It is well organized and illustrated.

It is of interest as a new contribution in studies of peculiarities of the structure and grain boundaries in submicrocrystalline materials obtained by severe plastic deformation. It is also recommended for materials science researchers dealing with the structure of various materials subjected to severe plastic deformation and with Nb-base materials such as heavily-deformed high-strength Cu-Nb composites.

(V.V. Kondratiev)
Date: March 20, 2013.

In: Niobium

Editors: M. Segers and Th. Peeters

ISBN: 978-1-62808-257-9

© 2013 Nova Science Publishers, Inc.

Chapter 2

CRYSTAL STRUCTURE OF ORDERED CARBIDE PHASES AND REVISED SEQUENCE OF PHASE TRANSFORMATIONS ASSOCIATED WITH ORDERING OF STRONGLY NONSTOICHIOMETRIC CARBIDES OF GROUP V TRANSITION METALS

A. I. Gusev and A. S. Kurlov

Institute of Solid State Chemistry,
Ural Division of the Russian Academy of Sciences,
Ekaterinburg, Russia

ABSTRACT

Nonstoichiometric carbides are unique entities for exploring atomic ordering. These compounds display a large variety of superstructures with different symmetry. The effect of ordering on all properties of nonstoichiometric carbides is large and comparable with variation of properties in the homogeneity interval of a disordered compound. In the present paper the data on order-disorder phase transformations in strongly nonstoichiometric carbides of Group V transition metals are reviewed. The main ordered phase of these compounds is the superstructure of M_6C_5 type. Taking into consideration recent findings we showed that trigonal (space group $P3_112$) and monoclinic (space groups $C2/m$ and

$C2/c$) M_6C_5 superstructures are formed as a result of ordering of nonstoichiometric cubic ($B1$ structure) carbides MC_y of Group V transition d metals. Symmetry analysis of the monoclinic and trigonal M_6C_5 superstructures is performed, $MC_y \to M_6C_5$ disorder-order transition channels are found, and the distribution functions of carbon atoms in M_6C_5 superstructures are calculated. It is shown that with a decrease in temperature, two physically admissible sequences of transformaions connected with the formation of M_6C_5 phases are possible in MC_y carbides of Group V transition metals. The first sequence MC_y (space group $Fm\bar{3}m$) $\to M_6C_5$ (space group $C2/m$) $\to M_6C_5$ (space group $C2/c$) includes disorder-order and order-order transformations. The second sequence involves disorder-order transformation MC_y (space group $Fm\bar{3}m$) $\to M_6C_5$ (space group $P3_112$) and a polymorphous transformation M_6C_5 (space group $P3_112$) $\to M_6C_5$ (space group $C2/c$).

Keywords: Nonstoichiometry; Carbide; Structural vacancy; Ordering; Superstructure; Symmetry analysis; Sequence of transformations

1. INTRODUCTION

Cubic carbides of transition d metals, especially carbides of niobium, vanadium and tantalum are used widely for a manufacture of constructional and functional materials. The principal application for carbides is as the major constituent in "cemented carbide" cutting tools. The term "cemented carbide" or hardmetals refers to a fourth to sixth group carbide bonded together in a metal matrix, usually cobalt. WC is the most important carbide for this application. Such carbides as NbC, TaC and VC used for this purpose usually as alloying additions to WC. Cobalt in the amounts 5-30 wt.% is added to increase the toughness of the tool bit without greatly reducing the hardness and it is also added because of the inability of carbides to self-sinter except under the application of very high temperatures.

Carbides have a unique set of properties necessary for use as a cutting tool. Property requirements include great hardness and wear resistance, good thermal shock resistance and thermal conductivity, good oxidation resistance, and compatibility of the carbide particles with the cobalt binder. Good thermal shock resistance and thermal conductivity are necessary to remove local heat generated at cutting surfaces. Since carbides are metallic conductors, this property is assured because of the rapid heat conduction by the free electrons.

Compatibility of carbides with the binder materials, usually cobalt, is also excellent.

Additions of NbC, TaC and VC carbides cause important changes in the properties of hardmetals. For example, additions of TaC and NbC raise the melting temperature of the carbide solution and increase the oxidation resistance.

Besides their use as cutting tools, the hardmetals are used extensively as spikes for snow tires, as wear resistant parts in wire drawing, extrusion and pressing dies, as drilling tools in mining, and as wear resistant surfaces in many types of machines and instruments.

In high-temperature applications, carbides are used either as self-bonded sintered parts or as a constituent in a sintered composite consisting of carbide particles bonded together with Co, Mo, or W. Typical parts include rocket nozzles and jet engine components. These materials are used where normal superalloys are no longer applicable because the required operating temperatures are too high. Self-bonded carbides such as NbC_y and VC_y alloys have great strengths up to about 2000 K and can be used as high temperature structural materials.

In refractory alloys of Nb, Ta, and V, carbides are used as a minor constituent to act as dispersion strengtheners. With increasing temperature, the solubility of carbon increases in these metals and therefore precipitation hardening is possible by controlled heat treatments. Fine precipitates of monoclinic (space group $C2/m$) ordered carbides V_6C_5 and Nb_6C_5 are detected in high-vanadium cast iron.

Niobium carbide is a frequent intentional product in microalloyed steels due to its extremely low solubility product in austenite, the lowest of all the refractory metal carbides. This means that micrometre-sized precipitates of NbC are virtually insoluble in steels at all processing temperatures and their location at grain boundaries helps prevent excessive grain growth in these steels. This is of enormous benefit, and the cornerstone of microalloyed steels, because it is their uniform, very fine grain size that ensures both toughness and strength. Also niobium carbide can be used as refractory coatings in nuclear reactors.

Another use for niobium alloys and niobium carbides is in superconducting devices. Alloy Nb_3Sn or niobium carbide NbC are used as materials for thin-film miniature superconducting solenoids, Josephson junctions, and bolometers. These materials are particularly promising for Josephson junctions; their refractory nature and corrosion resistance results in little chemical diffusion and hence decay of the junction with time.

Cubic carbides along with nitrides, oxides, and borides MX_y (X = C, N, O, B; $1/2 < y \leq 1$) of transition d metals are strongly nonstoichiometric interstitial compounds [1]. In MX_y compounds with the basic cubic (space group $Fm\bar{3}m$) structure $B1$, interstitial atoms X are located in metal sublattice octahedral interstitial sites forming of a face-centered cubic (fcc) nonmetallic sublattice. Depending on the relative content, y, the interstitial atoms X can occupy all or only some of the interstitial sites. The unoccupied interstitial sites are called structural vacancies $\square$. In nonstoichiometric compounds, the structural vacancies and interstitial atoms form a substitutional solid solution in the nonmetallic sublattice. The concentration of structural vacancies at the lower boundary of the homogeneity interval of MX_y nonstoichiometric compounds can vary from 0 to 30 at % or higher. A high concentration of structural vacancies is a prerequisite for the atomic-vacancy ordering of strongly nonstoichiometric compounds MX_y ($MX_y\square_{1-y}$).

As a result of ordering of nonmetal interstitial atoms X and structural vacancies $\square$ in the nonmetallic sublattice of cubic phases MX_y ($MX_y\square_{1-y}$) with the $B1$ basic structure, a large number of $M_{2t}X_{2t-1}$ ($M_{2t}X_{2t-1}\square$) superstructures are formed, where $t = 1, 1.5, 2, 3$, and 4. In the MX_y phases with a very small content of X atoms and very large content of vacancies $\square$, i.e., for $y < 1/2$, $M_{2t}X\square_{2t-1}$ superstructures are formed. At $y > 1/2$, inverse superstructures $M_{2t}X_{2t-1}\square$ with the same symmetry as that of $M_{2t}X\square_{2t-1}$ superstructures can be formed [2]. For example, Pd_6B (M_6X) superstructure formed in the $PdB_{0.16-0.18}$ solid solution [3] has a corresponding inverse superstructure M_6X_5. Here, inversion is treated as the inversion of the occupation of octahedral interstices: a vacant interstice is replaced by an interstice occupied by a nonmetal interstitial atom, and vice versa.

A particularly large body of experimental and theoretical data have been obtained for M_6C_5 ($M_6C_5\square$) superstructures with $t = 3$ [4]. In strongly nonstoichiometric cubic (space group $Fm\bar{3}m$) carbides MC_y of Group V transition d metals (M = V, Nb, Ta) with a relative carbon content $0.79 \leq y \leq 0.88$, the formation of M_6C_5 superstructures with different symmetries and distributions of carbon atoms and $\square$ vacancies over lattice sites has been experimentally observed at temperatures below 1300 K. For example, three ordered phases V_6C_5 with different symmetry has been found by different authors in vanadium carbide VC_y with the same carbon content. So, a question arises: what is a sequence of formation of M_6C_5 superstructures?

A possible sequence of phase transformations associated with the formation of M_6C_5 superstructures in nonstoichiometric carbides MC_y was discussed earlier in [5] where X-ray and neutron diffraction investigations of ordered phases of nonstoichiometric vanadium, niobium and tantalum carbides were performed. Recently, refined structures with new space groups have been proposed for ordered M_6C_5 phases.

2. ORDERED CARBIDES OF NIOBIUM, VANADIUM AND TANTALUM

Let us consider in more detail the available literature data on the crystal structure of ordered carbide M_6C_5 phases which form in cubic nonstoichiometric carbides MC_y of niobium, vanadium and tantalum which have the $B1$ structure and the basic disordered lattice constant a_{B1}.

It was revealed using electron diffraction, electron microscopy, and NMR spectroscopy that the axial ordered trigonal phase V_6C_5 is formed in the vanadium carbide $VC_{0.84}$ [6-8]. According to its symmetry, this superstructure belongs to the space group $P3_1$ or the enantiomorphic space group $P3_2$. More recently, Karimov et al. [9, 10] confirmed the formation of the ordered trigonal phase V_6C_5 with the use of structural neutron diffraction.

According to [9, 10], the refined trigonal structure of V_6C_5 phase belongs to the space group $P3_112$. Lipatnikov et al. [11] performed an X-ray diffraction analysis of the V_6C_5 phase and suggested that it should have a trigonal structure. Khaenko et al. [12, 13] carried out X-ray diffraction studies of annealed $VC_{0.82}$ crystallites and also revealed the trigonal (space groups $P3_112$ or $P3_121$) ordered phase V_6C_5. However, the unit cell proposed in [12, 13] is very large ($\mathbf{a}_{tr} = 3\langle 0\bar{1}\bar{1}\rangle_{B1}$, $\mathbf{b}_{tr} = 3\langle 110\rangle_{B1}$, $\mathbf{c}_{tr} = 12\langle 1\bar{1}1\rangle_{B1}$) and has not been confirmed in other studies. Lately, Cenzual et al. [14] revised the original experimental data [6] on the crystal structure of ordered nonstoichiometric vanadium carbide and showed that the trigonal superstructure V_6C_5 belongs to the space group $P3_112$ rather than to the space group $P3_1$. The lattice constants of the trigonal (space group $P3_112$) unit cell of V_6C_5 superstructure are $a = 0.509$ и $c = 1.44$ nm.

Almost simultaneously with the investigations performed in [6-8], the structure of nonstoichiometric vanadium carbide was examined by electron diffraction [15-18]. Billingham et al. [15] studied the crystal structure of ordered nonstoichiometric vanadium carbide in the composition range from $VC_{0.77}$ to $VC_{0.85}$. Instead of the trigonal superstructure, they found a base-centered monoclinic (space group $C2$ ($B2$)) superstructure V_6C_5 and demonstrated that almost all electron diffraction patterns described in [6] can be explained in terms of the monoclinic symmetry of the V_6C_5 phase rather than by the trigonal symmetry. Having analyzed the sequence of alternation of layers formed by carbon and vacancies in the monoclinic (space group $C2$ ($B2$)) and trigonal (space group $P3_1$) superstructures, Billingham et al. [15] noted that one more monoclinic superstructure with space group $C2/m$ ($B2/m$) is possible.

According to Lewis et al. [17], the domains of the ordered phase of the $VC_{0.83}$ carbide are characterized by a "double" structure, which includes monoclinic and trigonal atomic packings.

The structure of annealed carbides $VC_{0.78}$, $VC_{0.80}$, $VC_{0.84}$, and $VC_{0.86}$ was investigated using electron diffraction and electron microscopy by Hiraga [18]. According to [18], the distribution of carbon atoms and vacancies in the $VC_{0.84}$ carbide corresponds to a greater extent to the monoclinic (space group $C2$) model of the V_6C_5 phase [15] than to the trigonal model [6]. Moreover, for the V_6C_5 superstructure, Hiraga [18] proposed an orthorhombic unit cell with the parameters $a = (\sqrt{6}/2)a_{B1}$ and $b = (\sqrt{18}/2)a_{B1}$ equal to the corresponding parameters of the monoclinic (space group $C2$) unit cell and with the axis $c = (4\sqrt{3})a_{B1}$ parallel to the $[111]_{B1}$ direction. For ordered carbides VC_y ($y > 0.84$), a long-period superstructure $V_{44}C_{37}$ which is formed in the composition range between V_6C_5 and V_8C_7 phases was proposed [18].

The existence of the ordered monoclinic (space group $C2$) phase V_6C_5 was confirmed by the neutron diffraction investigation [19].

Kesri and Hamar_Thibault [20] studied the Fe-V-C system in the eutectic region between austenite and vanadium carbide and established that the solidification of cast iron containing 3.2 wt % C and 9.0 wt % V is accompanied by the precipitation of the trigonal vanadium carbide V_6C_5. Upon solidification of cast iron with less than 2 wt % C and more than 10 wt % V, the vanadium carbide has the monoclinic (space group $C2/m$) structure V_6C_5 at the centre of eutectic grains and the ordered cubic (space group $P4_332$) phase V_8C_7 surrounded by the Fe_3C cementite shell in the outer part of the grains. According to [20-22], the structure of the monoclinic phase V_6C_5 is similar to that of the ordered monoclinic (space group $C2/m$ ($C12/m1$)) niobium carbide Nb_6C_5 [23-25].

X-ray diffraction study of the crystal structure of annealed vanadium carbides $VC_{0.83}$ and $VC_{0.79}$ was performed in [5, 26, 27]. It demonstrated that the first phase formed as a result of ordering of these carbides can be both the monoclinic (space group $C2/m$) and the trigonal (space group $P3_1$) V_6C_5 superstructure. Indeed, the convergences between experimental and theoretical

X-ray diffraction patterns calculated for both structural models of the V_6C_5 phase are rather close to each other [5]. It should be noted that only the trigonal (space group $P3_1$) model of the V_6C_5 superstructure was considered in [5, 26, 27] while the model with space group $P3_112$ was not discussed.

The experimental data on nonstoichiometric niobium carbide NbC_y superstructures are less numerous than for the vanadium carbide. Ordering in the cubic (space group $Fm\bar{3}m$) niobium carbide NbC_y was examined using electron [16, 17, 20, 28] and neutron [23-25, 29-31] diffraction. It was found that the ordered Nb_6C_5 phase is formed over a wide range of compositions in the vicinity of the $NbC_{0.83}$ carbide. In [16, 17, 20, 29, 30], the observed superstructure reflections were described within the trigonal structure model similar to that used for the V_6C_5 superstructure [6]. In addition to neutron diffraction, Landesman et al. [29] studied the Nb_6C_5 phase by X-ray diffraction ($CuK\beta$-radiation) and experimentally revealed the trigonal splitting of the $(440)_{B1}$ structure reflection, which indicates a trigonal distortion of the cubic metal sublattice.

The neutron diffraction investigation [30] of the annealed single crystal $NbC_{0.83}$ confirmed the trigonal structure of the Nb_6C_5 phase and revealed a noticeable displacement of niobium and carbon atoms from all positions of the ideal structure, as well as the presence of carbon atoms with a probability of 0.52 at the lattice sites that should be completely vacant in the ideal trigonal superstructure Nb_6C_5. Although it was demonstrated in [29, 30] that the observed superstructure reflections do not correspond to the monoclinic superstructure Nb_6C_5 with space group $C2$, no alternative monoclinic (space group $C2/m$) structural model was discussed.

In [23-25, 31], the trigonal and two monoclinic models of the crystal structure of the ordered niobium carbide phase were analyzed. With the use of neutron and X-ray diffraction data, it was inferred that the ordered monoclinic (space group $C2/m$) phase Nb_6C_5 is formed in nonstoichiometric niobium carbide in the composition range from $NbC_{0.81}$ to $NbC_{0.88}$ upon annealing at temperatures below 1350 K. The formation of monoclinic (space group $C2/m$) ordered phase Nb_6C_5 was confirmed by NMR method in [32] and neutron diffraction in [5].

The X-ray diffraction study of the ordered niobium monocarbide [33, 34] showed the presence of the Nb_6C_5 phase with a hypothetical trigonal symmetry. More recently, the same authors [35] concluded that the Nb_6C_5 phase is monoclinic and belongs to space group $C2/m$.

Among the vanadium, niobium, and tantalum carbides, the cubic tantalum carbide TaC_y is the most complex object for the study of ordering. Experimental data on the ordering of the TaC_y carbides are almost absent.

The electron diffraction study of the $TaC_{0.83}$ carbide [28] revealed diffraction bands with the geometry corresponding to M_6C_5-type ordering with a very low degree of order. According to the neutron diffraction data [36-38], ordering of the TaC_y carbide leads to the formation of an incommensurate ordered structure similar to the M_6C_5-type structure. Thermodynamic calculations of the disorder-order transformations [1, 2, 39-41] showed that the M_6C_5-type superstructure can be the only ordered phase of the TaC_y carbide.

The superstructure M_6X_5, which is inverse to the Pd_6B superstructure, is discussed in [42, 43]. It is shown [42, 43] that the distribution of interstitial atoms and structural vacancies in the monoclinic superstructure M_6X_5 with space groups $C2$ and $C2/c$ is completely identical and therefore the monoclinic superstructure is described more correctly in space group $C2/c$ rather than in space group $C2$.

Thus, several different M_6C_5 superstructures for the ordered nonstoichiometric vanadium, niobium, and tantalum carbide have been described in the literature: trigonal (space groups $P3_1$ or $P3_112$) and monoclinic (space groups $C2/m$ and also $C2$ or $C2/c$) structures.

In the present work, with allowance for new data, we carried out a symmetry analysis of a possible sequence of phase transformations associated with the formation of M_6C_5 superstructures in nonstoichiometric carbides MC_y.

3. TRIGONAL M_6C_5 SUPERSTRUCTURE

To perform symmetry analysis and calculate the distribution function of carbon atoms in the M_6C_5 superstructures under consideration, it is necessary to consider the reciprocal lattice of these superstructures and determine the channel of the disorder-order structural phase transition MC_y - M_6C_5. The basis vectors $\mathbf{b}_i^*$ of the reciprocal lattice are determined via translation vectors $\mathbf{a}_i$ of the unit cell by the ordinary formula

$$\mathbf{b}_i^* = 2\pi \frac{\mathbf{a}_j \times \mathbf{a}_k}{\mathbf{a}_1(\mathbf{a}_2 \times \mathbf{a}_3)}, \tag{1}$$

where $i, j, k = 1, 2, 3$.

The trigonal (space groups $P3_112$ and $P3_1$) unit cells of the M_6C_5 superstructure are shown in Figure 1. The coordinates of atoms and vacancies in the perfect trigonal (space groups $P3_112$ and $P3_1$) M_6C_5 superstructure are given in Table 1.

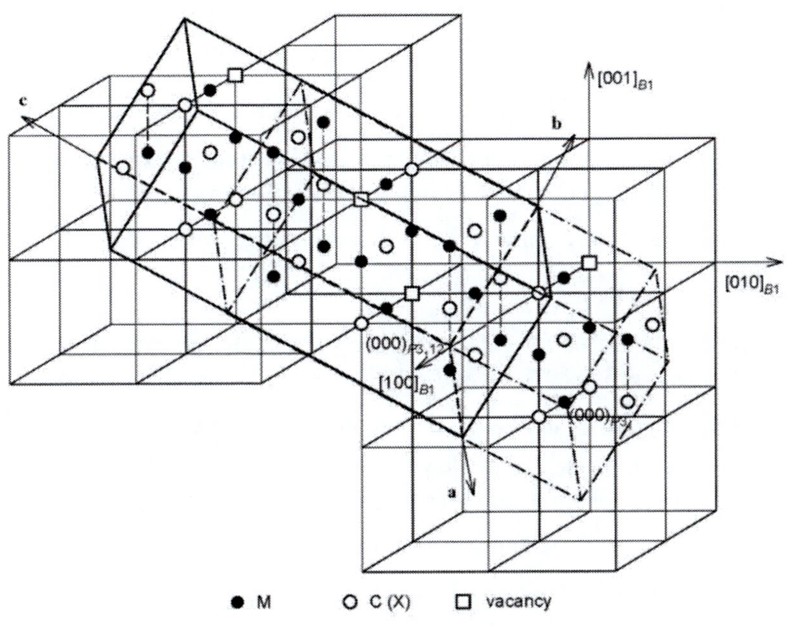

Figure 1. Position of the trigonal (space groups $P3_112$ and $P3_1$) unit cells of the M_6C_5 superstructure in a lattice with the $B1$ structure. Contours of these unit cells are shown by solid lines and chain lines, respectively. The origin of coordinates $(000)_{P3_112}$ of the trigonal (space group $P3_112$) unit cell has the cubic coordinates $\left(\frac{2}{3}\,\frac{\bar{5}}{6}\,\frac{\bar{1}}{2}\right)_{B1}$, the origin of coordinates $(000)_{P3_1}$ of the trigonal (space group $P3_1$) unit cell has the cubic coordinates $\left(0\,\frac{\bar{1}}{6}\,\frac{7}{6}\right)_{B1}$. The trigonal (space group $P3_112$) unit cell is shifted relative to the trigonal (space group $P3_1$) unit cell by one third of $\mathbf{c}_{tr}$ axis length, i. e. by the vector $\mathbf{c}_{tr}/3 = \frac{2}{3}\langle 1\bar{1}1\rangle_{B1}$. (●) metal atoms M, (○) interstitial carbon atoms C, and (□) vacancies.

Both unit cells have the same translation vectors which in the basic lattice with the $B1$ structure are equal to $\mathbf{a}_1 = \mathbf{a}_{tr} = \frac{1}{2}\langle 21\bar{1}\rangle_{B1}$, $\mathbf{a}_2 = \mathbf{b}_{tr} = \frac{1}{2}\langle\bar{1}12\rangle_{B1}$, and $\mathbf{a}_3 = \mathbf{c}_{tr} = 2\langle 1\bar{1}1\rangle_{B1}$. The origin of coordinates $(000)_{P3_1}$ of the trigonal (space group $P3_1$) unit cell has cubic coordinates $(0\,\frac{\bar{1}}{6}\,\frac{\bar{7}}{6})_{B1}$. The unit cell of the M_6C_5 superstructure with space group $P3_112$ is shifted relative to the trigonal (space group $P3_1$) unit cell by one third of $\mathbf{c}_{tr}$ axis length, i. e. by the vector $\mathbf{c}_{tr}/3 = \frac{2}{3}\langle 1\bar{1}1\rangle_{B1}$. The origin of the coordinates $(000)_{P3_112}$ of the trigonal (space group $P3_112$) unit cell has cubic coordinates $(\frac{2}{3}\,\frac{\bar{5}}{6}\,\frac{\bar{1}}{2})_{B1}$. The unit cells of the trigonal superstructures involve three formula units M_6C_5. In accordance with Eq.(1) the basis vectors of the reciprocal lattice of the trigonal (space groups $P3_112$ and $P3_1$) superstructures M_6C_5 are $\mathbf{a}_{tr}^* = \frac{2}{3}\langle 110\rangle$, $\mathbf{b}_{tr}^* = \frac{2}{3}\langle 011\rangle$ and $\mathbf{c}_{tr}^* = \frac{1}{6}\langle 1\bar{1}1\rangle$.

The translation of all superstructure sites of the reciprocal lattices of the trigonal (space groups $P3_112$ and $P3_1$) superstructures M_6C_5 indicates that the first Brillouin zone of the basic disordered fcc lattice contains 13 rays $\mathbf{k}_s^{(j)}$ of the three stars $\{\mathbf{k}_9\}$, $\{\mathbf{k}_4\}$, and $\{\mathbf{k}_3\}$ (Table 2).

Hereinafter, the numbering and description of the wave-vector stars $\{\mathbf{k}_s\}$ and their rays $\mathbf{k}_s^{(j)}$ are given according to [1, 2, 44, 45]; $\mathbf{b}_1 = \langle\bar{1}1\,1\rangle$, $\mathbf{b}_2 = \langle 1\bar{1}1\rangle$, and $\mathbf{b}_3 = \langle 1\,1\bar{1}\rangle$ are the structure vectors of the reciprocal lattice of the basic fcc lattice in $2\pi/a$ units. These 13 nonequivalent superstructure vectors are involved in the phase transition channel associated with the formation of the considered trigonal M_6C_5 superstructures.

The structure of ordered phases is conveniently described using the distribution function $n(\mathbf{r})$, which represents the probability of finding an atom of a particular species at the site $\mathbf{r} = (x_I, y_I, z_I)$ in the ordering Ising lattice. In the MC_y carbides with the $B1$ basic structure, the nonmetallic face-centered cubic sublattice is the Ising lattice, in which atomic-vacancy ordering can take place. The deviation of the probability $n(\mathbf{r})$ from its value in case of a disordered (statistical) distribution can be represented as a superposition of several plane concentration waves [46].

Table 1. Trigonal M_6C_5-type superstructures: space group No 151: $P3_112$ (D_3^3) and space group No 144: $P3_1$ (C_3^2).

$$Z = 3,\ \mathbf{a_{tr}} = \tfrac{1}{2}\langle 21\bar{1}\rangle_{B1},\ \mathbf{b_{tr}} = \tfrac{1}{2}\langle \bar{1}12\rangle_{B1},\ \mathbf{c_{tr}} = 2\langle 1\bar{1}1\rangle_{B1}$$

Atom	Space group $P3_112$				Space group $P3_1$				Atomic coordinates in the basic disordered $B1$ structure (according to Figure 1)			Probabilities of detecting C atom at different positions of M_6C_5 superstructures, i. e., values of the distribution function $n(x_I, y_I, z_I)$ (3) for C atoms
	position and multiplicity	atomic coordinates in the ideal ordered structure			position and multiplicity	atomic coordinates in the ideal ordered structure						
		x/a_{tr}	y/b_{tr}	z/c_{tr}		x/a_{tr}	y/b_{tr}	z/c_{tr}	x/a_{B1}	y/a_{B1}	z/a_{B1}	
C1 (vacancy)	Special 3(b)	1/9	2/9	1/6	General 3(a)	1/9	2/9	1/2	1	-1	0	$n_1 = y - \eta_9/6 - \eta_4/3 - \eta_3/3$
C2	Special 3(b)	4/9	8/9	1/6	General 3(a)	4/9	8/9	1/2	1	-1/2	1/2	$n_3 = y - \eta_9/6 + \eta_4/6 + \eta_3/6$
C3	Special 3(b)	7/9	5/9	1/6	General 3(a)	7/9	5/9	1/2	3/2	-1/2	0	$n_3 = y - \eta_9/6 + \eta_4/6 + \eta_3/6$
C4	Special 3(a)	1/9	8/9	1/3	General 3(a)	1/9	8/9	2/3	1	-1	1	$n_2 = y + \eta_9/6 - \eta_4/3 + \eta_3/3$
C5	Special 3(a)	4/9	5/9	1/3	General 3(a)	4/9	5/9	2/3	3/2	-1	1/2	$n_4 = y + \eta_9/6 + \eta_4/6 - \eta_3/6$
C6	Special 3(a)	7/9	2/9	1/3	General 3(a)	7/9	2/9	2/3	2	-1	0	$n_4 = y + \eta_9/6 + \eta_4/6 - \eta_3/6$
M1	General 6(c)	4/9	5/9	1/12	General 3(a)	4/9	5/9	5/12	1	-1/2	0	
		4/9	5/9	7/12	General 3(a)	4/9	5/9	11/12	2	-3/2	1	
M2	General 6(c)	1/9	8/9	1/12	General 3(a)	1/9	8/9	5/12	1/2	-1/2	1/2	
		1/9	8/9	7/12	General 3(a)	1/9	8/9	11/12	3/2	-3/2	3/2	
M3	General 6(c)	7/9	2/9	1/12	General 3(a)	7/9	2/9	5/12	3/2	-1/2	-1/2	
		7/9	2/9	7/12	General 3(a)	7/9	2/9	11/12	5/2	-3/2	1/2	

**Table 2. $MC_y - M_6C_5$ disorder-order phase transition channel and parameters
of the distribution function
$n(x_I, y_I, z_I)$ (3) describing the trigonal (space groups $P3_112$ and $P3_1$) M_6C_5 superstructures**

Space groups	Disorder-order phase transition channel		Parameters of the distribution function (3)	
	wave-vector star $\{k_s\}$	rays $\mathbf{k}_s^{(j)}$ of the star $\{k_s\}$	γ_s	$\varphi_s^{(j)}$
$P3_112,$ $P3_1$	$\{\mathbf{k}_9\}$	$\mathbf{k}_9^{(3)} = \mathbf{b}_2/2 = 3\,\mathbf{c}_{tr}^* = \left\langle \frac{1}{2}\,\frac{\bar{1}}{2}\,\frac{1}{2} \right\rangle,$	$\gamma_9 = 1/6$	$\varphi_9^{(3)} = \pi$
	$\{\mathbf{k}_4\}$	$\mathbf{k}_4^{(1)} = (\mathbf{b}_1 + \mathbf{b}_2 + 2\mathbf{b}_3)/3 = \mathbf{a}_{tr}^* = \left\langle \frac{2}{3}\,\frac{2}{3}\,0 \right\rangle,$	$\gamma_4 = \sqrt{3}/18$	$\varphi_4^{(1)} = 7\pi/6$
	$\{\mathbf{k}_4\}$	$\mathbf{k}_4^{(2)} = -\mathbf{k}_4^{(1)},$	$\gamma_4 = \sqrt{3}/18$	$\varphi_4^{(2)} = -7\pi/6$
	$\{\mathbf{k}_4\}$	$\mathbf{k}_4^{(7)} = (\mathbf{b}_3 - \mathbf{b}_1)/3 = \mathbf{a}_{tr}^* - \mathbf{b}_{tr}^* = \left\langle \frac{2}{3}\,0\,\frac{\bar{2}}{3} \right\rangle,$	$\gamma_4 = \sqrt{3}/18$	$\varphi_4^{(7)} = 3\pi/2$
	$\{\mathbf{k}_4\}$	$\mathbf{k}_4^{(8)} = -\mathbf{k}_4^{(7)},$	$\gamma_4 = \sqrt{3}/18$	$\varphi_4^{(8)} = -3\pi/2$
	$\{\mathbf{k}_4\}$	$\mathbf{k}_4^{(9)} = (2\mathbf{b}_1 + \mathbf{b}_2 + \mathbf{b}_3)/3 = \mathbf{b}_{tr}^* = \left\langle 0\,\frac{2}{3}\,\frac{2}{3} \right\rangle,$	$\gamma_4 = \sqrt{3}/18$	$\varphi_4^{(9)} = 5\pi/6$
	$\{\mathbf{k}_4\}$	$\mathbf{k}_4^{(10)} = -\mathbf{k}_4^{(9)},$	$\gamma_4 = \sqrt{3}/18$	$\varphi_4^{(10)} = -5\pi/6$
	$\{\mathbf{k}_3\}$	$\mathbf{k}_3^{(3)} = -(4\mathbf{b}_1 + \mathbf{b}_2 + 2\mathbf{b}_3)/6 = -\mathbf{b}_{tr}^* + \mathbf{c}_{tr}^* = \left\langle \frac{1}{6}\,\frac{\bar{5}}{6}\,\frac{\bar{1}}{2} \right\rangle,$	$\gamma_3 = \sqrt{3}/18$	$\varphi_3^{(3)} = 7\pi/6$
	$\{\mathbf{k}_3\}$	$\mathbf{k}_3^{(4)} = -\mathbf{k}_3^{(3)},$	$\gamma_3 = \sqrt{3}/18$	$\varphi_3^{(4)} = -7\pi/6$
	$\{\mathbf{k}_3\}$	$\mathbf{k}_3^{(9)} = (2\mathbf{b}_1 + 3\mathbf{b}_2 + 4\mathbf{b}_3)/6 = \mathbf{a}_{tr}^* + \mathbf{c}_{tr}^* = \left\langle \frac{5}{6}\,\frac{1}{2}\,\frac{1}{6} \right\rangle,$	$\gamma_3 = \sqrt{3}/18$	$\varphi_3^{(9)} = \pi/2$
	$\{\mathbf{k}_3\}$	$\mathbf{k}_3^{(10)} = -\mathbf{k}_3^{(9)},$	$\gamma_3 = \sqrt{3}/18$	$\varphi_3^{(10)} = -\pi/2$
	$\{\mathbf{k}_3\}$	$\mathbf{k}_3^{(23)} = (2\mathbf{b}_1 + \mathbf{b}_2 - 2\mathbf{b}_3)/6 = -\mathbf{a}_{tr}^* + \mathbf{b}_{tr}^* + \mathbf{c}_{tr}^* = \left\langle \frac{\bar{1}}{2}\,\frac{\bar{1}}{6}\,\frac{5}{6} \right\rangle,$	$\gamma_3 = \sqrt{3}/18$	$\varphi_3^{(23)} = -5\pi/6$
	$\{\mathbf{k}_3\}$	$\mathbf{k}_3^{(24)} = -\mathbf{k}_3^{(23)}$	$\gamma_3 = \sqrt{3}/18$	$\varphi_3^{(24)} = 5\pi/6$

The wave vectors of these waves are the superstructure vectors forming a disorder-order transition channel [1, 2].

In the method of static concentration waves [46], the distribution function $n(\mathbf{r})$ has the form

$$n(\mathbf{r}) = y + \frac{1}{2}\sum_{s}\sum_{j\in s}\eta_s\gamma_s[\exp(i\varphi_s^{(j)})\exp(i\mathbf{k}_s^{(j)}\mathbf{r}) + \exp(-i\varphi_s^{(j)})\exp(-i\mathbf{k}_s^{(j)}\mathbf{r})], \qquad (2)$$

where y is the fraction of sites occupied by atoms of a particular species in the ordering sublattice;

$$\frac{1}{2}\eta_s\gamma_s[\exp(i\varphi_s^{(j)})\exp(i\mathbf{k}_s^{(j)}\mathbf{r}) + \exp(-i\varphi_s^{(j)})\exp(-i\mathbf{k}_s^{(j)}\mathbf{r})] \equiv \Delta(\mathbf{k}_s^{(j)},\mathbf{r})$$

is the standing plane static concentration wave generated by the superstructure vector $\mathbf{k}_s^{(j)}$ of the star $\{\mathbf{k}_s\}$; η_s is the long-range order parameter corresponding to the star $\{\mathbf{k}_s\}$; $\eta_s\gamma_s$ and $\varphi_s^{(j)}$ are the amplitude and the phase shift of the concentration wave, respectively.

The function $n(\mathbf{r})$ takes on the same value at the sites $\mathbf{r}$ located at crystallographically equivalent positions.

The total number of values taken by the distribution function is larger than the number of long-range order parameters by unity.

The summation in formula (2) should be performed only over nonequivalent superstructure vectors of the first Brillouin zone.

The identical disorder-order phase transition channel corresponding to the trigonal (space groups $P3_112$ and $P3_1$) superstructures M_6C_5 (see Table 2) means that these superstructures are described the same distribution function.

The method for calculating distribution function $n(\mathbf{r}) \equiv n(x_I, y_I, z_I)$ and its parameters $\eta_s\gamma_s$ and $\varphi_s^{(j)}$ is described in detail in review [4].

With allowance for the obtained values of $\eta_s\gamma_s$ and $\varphi_s^{(j)}$ (Table 2), the distribution function of carbon atoms in the trigonal (space groups $P3_112$ and $P3_1$) superstructures M_6C_5 depends on the three long-range order parameters η_9, η_4, and η_3, corresponding to the stars $\{\mathbf{k}_9\}$, $\{\mathbf{k}_4\}$, and $\{\mathbf{k}_3\}$, respectively, and has the form

$$n(x_I, y_I, z_I) = y - (\eta_9/6)\cos[\pi(x_I - y_I + z_I)] -$$

$$- (\eta_4/6)\{\cos[4\pi(x_I + y_I)/3] - (\sqrt{3}/3)\sin[4\pi(x_I + y_I)/3] -$$

$$- (2\sqrt{3}/3)\sin[4\pi(x_I - z_I)/3] + \cos[4\pi(y_I + z_I)/3] + (\sqrt{3}/3)\sin[4\pi(y_I + z_I)/3]\} -$$

$$- (\eta_3/6)\{\cos[\pi(x_I - 5y_I - 3z_I)/3] - (\sqrt{3}/3)\sin[\pi(x_I - 5y_I - 3z_I)/3] +$$

$$+ (2\sqrt{3}/3)\sin[\pi(5x_I + 3y_I + z_I)/3] + \cos[\pi(3x_I + y_I - 5z_I)/3] +$$

$$+ (\sqrt{3}/3)\sin[\pi(3x_I + y_I - 5z_I)/3]\}. \tag{3}$$

At all sites of the basic nonmetallic fcc sublattice, the distribution function (3), which describes the trigonal superstructures M_6C_5, takes on four different values, namely, $n_1 = y - \eta_9/6 - \eta_4/3 - \eta_3/3$, $n_2 = y + \eta_9/6 - \eta_4/3 + \eta_3/3$, $n_3 = y - \eta_9/6 + \eta_4/6 + \eta_3/6$, and $n_4 = y + \eta_9/6 + \eta_4/6 - \eta_3/6$ (Table 2). This means that the nonmetallic sublattice of the disordered nonstoichiometric carbide MC_y is divided into four nonequivalent sublattices as a result of the ordering considered. The probability of occupation of sites of the first sublattice by C atoms is equal to n_1, the probability of occupation of sites of the second sublattice is equal to n_2, etc. For the perfect M_6C_5 superstructures, the value of y in distribution function (3), i. e., the relative content of C atoms (or, in the general case, interstitial atoms X) is equal to 5/6.

The identity of the transition channel and the distribution function means that only one of the two trigonal models of the structure is valid for the M_6C_5 phase. The choice of the unit cell in the same lattice satisfying the conventional requirements (correspondence of the crystal symmetry, the maximal number of right angles in the cell, and the minimal volume of the cell) can be made in different ways and is ambiguous [47]. As a result, the same crystal is described in experimental studies in different ways.

In fact, the requirements on the choice of the unit cell boil down to that it have the highest possible symmetry. For the same unit cell volume, this means that the point symmetry group of the chosen unit cell must include the maximal number of symmetry elements (operations).

The point symmetry group 322 (D_3) of the trigonal (space group $P3_121$) M_6C_5 superstructure includes six symmetry elements (rotations) h_1, h_5, h_9, h_{13}, h_{17}, and h_{21}, while the point group 3 (C_3) of the trigonal (space group $P3_1$) M_6C_5 superstructure includes three symmetry elements h_1, h_5, and h_9 [1, 44, 45]. Thus, the trigonal (space group $P3_121$) model of the structure of the M_6C_5 phase is characterized by a higher symmetry as compared to the trigonal model of the M_6C_5 phase with space group $P3_1$. Therefore, the trigonal (space group

$P3_121$) model describes the crystal structure of the ordered phase M_6C_5 ($M_6C_5\square$) more adequately than the model with space group $P3_1$. This means that the trigonal (space group $P3_1$) M_6C_5 superstructures experimentally determined in [6-8, 11, 20] and described later in reviews [1, 2, 4, 40, 41] actually belong to space group $P3_121$.

4. MONOCLINIC M_6C_5 SUPERSTRUCTURE

4.1. Space Group $C2/c$ or $C2$?

Recently, the PdB_y solid solution of boron in fcc palladium was investigated [48, 49] using X-ray, neutron, and electron diffraction. The PdB_y disordered solid solution has a basic cubic $B1$ structure. In studying the PdB_y solid solution, Berger et al. [48, 49] used an analogy with ordered nonstoichiometric carbides M_6C_5 [1, 2, 4, 50] and revealed that the monoclinic (space group $C2/c$ ($C12/c1$)) superstructure Pd_6B (M_6X) is formed. The inverse M_6X_5 superstructure with the same space group corresponds to the monoclinic (space group $C2/c$) superstructure Pd_6B. For the inverse monoclinic (space group $C2/c$) superstructure M_6C_5 (M_6X_5), the disorder-order structural phase transition channel was determined in [42, 43]. It turned out that this channel is similar to the disorder-order transition channel in which the monoclinic superstructure M_6C_5 with space group $C2$ is formed. Let us consider whether the monoclinic superstructures M_6C_5 with space groups $C2/c$ and $C2$ differ physically. The upper part of Figure 2 shows the position of a unit cell of the monoclinic (space group $C2$) M_6C_5 superstructure as well as the contour of the unit cell of the monoclinic (space group $C2/c$) M_6C_5 phase. The arrangement of atoms and vacancies in the unit cell of the monoclinic (space group $C2/c$) phase M_6C_5 ($M_6C_5\square$) is shown in lower part of Figure 2. In accordance with Figure 2, the origin of coordinates $(0\ 0\ 0)_{C2/c}$ of this superstructure has the cubic coordinates $(\frac{\overline{1}}{2}\frac{1}{4}\frac{\overline{1}}{4})_{B1}$, i.e. it is displaced relative to the coordinate origin $(0\ 0\ 0)_{B1} \equiv (0\ 0\ 0)_{C2}$ of the monoclinic (space group $C2$) unit cell of the M_6C_5 superstructure by vector $\frac{1}{4}\langle 2\ 1\ \overline{1} \rangle_{B1}$. The coordinates of atoms and vacancies in the perfect monoclinic (space groups $C2/c$ и $C2$) superstructures M_6C_5 ($M_6C_5\square$) are given in Table 3. The unit cells of the monoclinic (space groups $C2/c$ и $C2$) superstructures involve four formula units M_6C_5 and have the same volume.

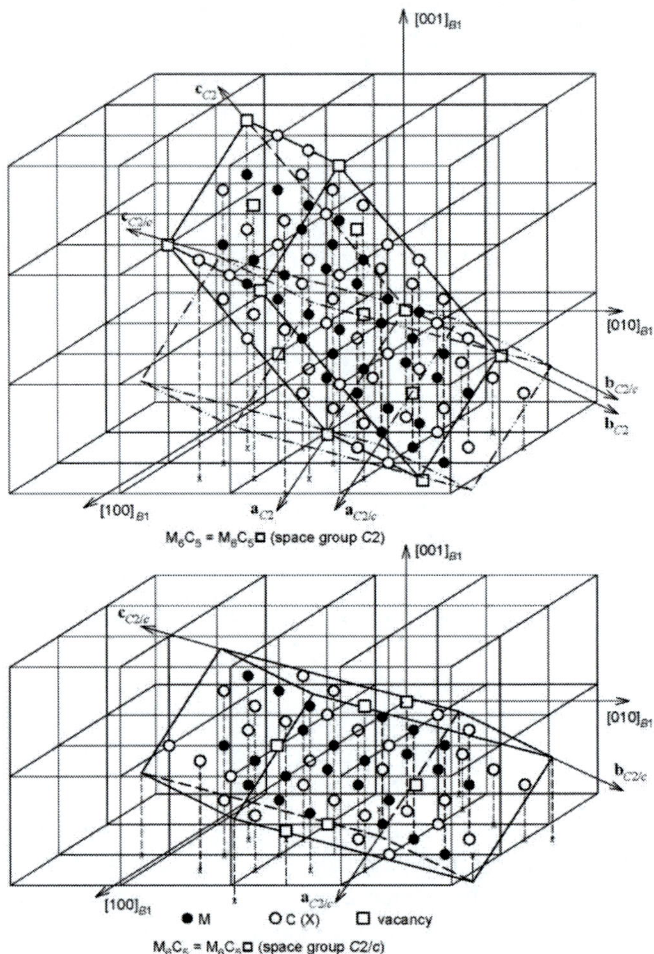

Figure 2. Position of the monoclinic (space groups $C2$ and $C2/c$) units cells of the M_6C_5 superstructure in a lattice with the $B1$ structure. Chain lines on an upper figure additionally show the contour of a unit cell of the M_6C_5 superstructure with monoclinic (space group $C2/c$) symmetry. The origin of coordinates $(000)_{C2/c}$ of the monoclinic (space group $C2/c$) unit cell has the cubic coordinates $\left(\frac{\overline{1}}{2}\frac{1}{4}\frac{\overline{1}}{4}\right)_{B1}$; i.e., it is displaced relative to the origin of coordinates $(000)_{B1} \equiv (000)_{C2}$ of the monoclinic (space group $C2$) unit cell by a vector $\frac{1}{4}\langle \overline{2}\,1\,\overline{1}\rangle_{B1}$. It is seen that different unit cells are chosen in the same ordered crystal structure. Vertical chain lines show projections ($\times$) of atoms and vacancies as well as the unit cell vertices onto the $(x\,y\,0)_{B1}$ plane. ($\bullet$) metal atoms M, ($\bigcirc$) interstitial carbon atoms C, and ($\square$) vacancies.

The translation vectors of the unit cell of the perfect monoclinic (space group $C2$) phase M_6C_5 in a basic lattice with the $B1$ structure has the form $\mathbf{a}_{C2} = \frac{1}{2}\langle 1\,\bar{1}\,\bar{2}\rangle_{B1}$, $\mathbf{b}_{C2} = \frac{1}{2}\langle 3\,3\,0\rangle_{B1}$, and $\mathbf{c}_{C2} = \langle 1\,\bar{1}\,2\rangle_{B1}$ while the translation vectors of the unit cell of the perfect monoclinic phase M_6C_5 with space group $C2/c$ are equal to $\mathbf{a}_{C2/c} = \frac{1}{2}\langle 1\,\bar{1}\,\bar{2}\rangle_{B1}$, $\mathbf{b}_{C2/c} = \frac{1}{2}\langle 3\,3\,0\rangle_{B1}$, and $\mathbf{c}_{C2/c} = \frac{1}{2}\langle 3\,\bar{3}\,2\rangle_{B1}$.

It is seen from Figure 2 that the translation vectors of the unit cell of the monoclinic (space group $C2/c$) phase M_6C_5 coincide in length and direction with the translation vectors of the monoclinic (space group $C2$) M_6C_5 phase or are their linear combinations: $\mathbf{a}_{C2/c} = \mathbf{a}_{C2}$, $\mathbf{b}_{C2/c} = \mathbf{b}_{C2}$, and $\mathbf{c}_{C2/c} = \mathbf{a}_{C2} + \mathbf{c}_{C2}$.

The volumes of the unit cells are the same. In accordance with Eq.(1), the basis vectors of the reciprocal lattice of the monoclinic (space group $C2$) superstructure M_6C_5 are $\mathbf{a}^*_{C2} = \frac{1}{2}\langle 1\,\bar{1}\,\bar{1}\rangle$, $\mathbf{b}^*_{C2} = \frac{1}{3}\langle \bar{1}\,\bar{1}\,0\rangle$, and $\mathbf{c}^*_{C2} = \frac{1}{4}\langle 1\,\bar{1}\,1\rangle$

In view of the invariance of determinants of formula (1) relative to addition and subtraction of the lines, the superstructure reciprocal lattice vectors of the monoclinic (space group $C2/c$) phase M_6C_5 also coincide or are combinations of the reciprocal lattice vectors of the monoclinic (space group $C2$) phase M_6C_5: $\mathbf{a}^*_{C2/c} = \mathbf{a}^*_{C2} - \mathbf{c}^*_{C2} = \frac{1}{4}\langle 1\,\bar{1}\,\bar{3}\rangle$, $\mathbf{b}^*_{C2/c} = \mathbf{b}^*_{C2} = \frac{1}{3}\langle \bar{1}\,\bar{1}\,0\rangle$, and $\mathbf{c}^*_{C2/c} = \mathbf{c}^*_{C2} = \frac{1}{4}\langle 1\,\bar{1}\,1\rangle$. It is clear that the monoclinic (space groups $C2/c$ and $C2$) phases M_6C_5 are formed by the same disorder-order transition channel.

The combinations and translations of the obtained superstructure reciprocal lattice vectors of the monoclinic (space groups $C2/c$ and $C2$) phases M_6C_5 show that the first Brillouin zone of the basic disordered fcc lattice contains nine rays $\mathbf{k}_s^{(j)}$ (Table 4) which belong to the Lifschitz star $\{\mathbf{k}_9\}$, non-Lifschitz stars $\{\mathbf{k}_4\}$ and $\{\mathbf{k}_3\}$, as well as to the non-Lifschitz star $\{\mathbf{k}_0\}$ of the common type.

These nine superstructure vectors enter the phase transition channel associated with the formation of the monoclinic (space groups $C2/c$ and $C2$) superstructures M_6C_5 under consideration.

Table 3. Monoclinic M_6C_5-type superstructures: space group No 15: $C2/c$ ($C12/c1$ (C_{2h}^6))

and space group No 5: $C2$ ($C121$ (C_2^3)).

$$Z = 4,\ \mathbf{a}_{C2/c} = \tfrac{1}{2}\langle 1\,\overline{1}\,\overline{2}\rangle_{B1},\ \mathbf{b}_{C2/c} = \tfrac{1}{2}\langle 3\,3\,0\rangle_{B1},\ \mathbf{c}_{C2/c} = \tfrac{1}{2}\langle 3\,\overline{3}\,2\rangle_{B1};\ \mathbf{a}_{C2} = \tfrac{1}{2}\langle 1\,\overline{1}\,\overline{2}\rangle_{B1},\ \mathbf{b}_{C2} = \tfrac{1}{2}\langle 3\,3\,0\rangle_{B1},\ \mathbf{c}_{C2} = \langle 1\,\overline{1}\,2\rangle_{B1}$$

Atom	Space group $C2/c$				Space group $C2$				Atomic coordinates in the basic disordered $B1$ structure (according to Figure 2)			Probabilities of detecting C atom at different positions of M_6C_5 superstructures, i. e., values of the distribution function $n(x_1, y_1, z_1)$ (4) for C atoms
	position and multiplicity	atomic coordinates in the ideal ordered structure			position and multiplicity	atomic coordinates in the ideal ordered structure						
		$x/a_{C2/c}$	$y/b_{C2/c}$	$z/c_{C2/c}$		x/a_{C2}	y/b_{C2}	z/c_{C2}	x/a_{B1}	y/a_{B1}	z/a_{B1}	
C1 (vacancy)	Special 4(e)	0	1/12	1/4	Special 2(a)	0	0	0	0	0	0	$n_1 = y - \eta_9/6 - \eta_4/12 - \eta_3/12 - \eta_0/2$
		1/2	5/12	3/4	Special 2(b)	1	1/3	1/2	3/2	-1/2	0	$n_1 = y - \eta_9/6 - \eta_4/12 - \eta_3/12 - \eta_0/2$
C2	Special 4(e)	0	5/12	1/4	Special 2(a)	0	1/3	0	1/2	1/2	0	$n_5 = y - \eta_9/6 - \eta_4/12 - \eta_3/12 + \eta_0/2$
		0	7/12	3/4	Special 2(b)	1/2	1/2	1/2	3/2	0	1/2	$n_5 = y - \eta_9/6 - \eta_4/12 - \eta_3/12 + \eta_0/2$
C3	Special 4(e)	0	3/4	1/4	Special 2(a)	0	2/3	0	1	1	0	$n_3 = y - \eta_9/6 + \eta_4/6 + \eta_3/6$
		0	1/4	3/4	Special 2(b)	1/2	1/6	1/2	1	-1/2	1/2	$n_3 = y - \eta_9/6 + \eta_4/6 + \eta_3/6$
C4	Special 4(c)	1/4	3/4	1/2	General 4(c)	1/2	2/3	1/4	3/2	1/2	0	$n_4 = y + \eta_9/6 + \eta_4/6 - \eta_3/6$
C5	General 8(f)	1/4	5/12	1/2	General 4(c)	1/2	1/3	1/4	1	0	0	$n_2 = y + \eta_9/6 - \eta_4/12 + \eta_3/12$
		1/4	1/12	1/2	General 4(c)	1/2	0	1/4	1/2	-1/2	0	$n_2 = y + \eta_9/6 - \eta_4/12 + \eta_3/12$
M1	General 8(f)	1/8	1/4	3/8	General 4(c)	1/4	1/6	1/8	1/2	0	0	
		1/8	3/4	7/8	General 4(c)	3/4	2/3	5/8	2	0	1/2	
M2	General 8(f)	5/8	5/12	3/8	General 4(c)	3/4	1/3	1/8	1	0	-1/2	
		3/8	7/12	5/8	General 4(c)	3/4	1/2	3/8	3/2	0	0	
M3	General 8(f)	1/8	7/12	3/8	General 4(c)	1/4	1/2	1/8	1	1/2	0	
		5/8	1/12	3/8	General 4(c)	3/4	0	1/8	1/2	-1/2	-1/2	

Table 4. $MC_y - M_6C_5$ disorder-order phase transition channels and parameters of the distribution function $n(x_I, y_I, z_I)$ (4) describing the monoclinic (space groups $C2/c$ and $C2$) M_6C_5 superstructures

Space groups	Disorder-order phase transition channel		Parameters of the distribution function (4)	
	wave-vector star $\{\mathbf{k_s}\}$	rays $\mathbf{k}_s^{(j)}$ of the star $\{\mathbf{k_s}\}$	γ_s	$\varphi_s^{(j)}$
$C2/c$, $C2$	$\{\mathbf{k}_9\}$	$\mathbf{k}_9^{(3)} = \mathbf{b}_2/2 = 2\mathbf{c}_{C2/c}^* = 2\mathbf{c}_{C2}^* = \left\langle \frac{1}{2}\,\frac{\bar{1}}{2}\,\frac{1}{2} \right\rangle,$	$\gamma_9 = 1/6$	$\varphi_9^{(3)} = \pi$
	$\{\mathbf{k}_4\}$	$\mathbf{k}_4^{(1)} = (\mathbf{b}_1 + \mathbf{b}_2 + 2\mathbf{b}_3)/3 = -2\mathbf{b}_{C2/c}^* = -2\mathbf{b}_{C2}^* = \left\langle \frac{2}{3}\,\frac{2}{3}\,0 \right\rangle,$	$\gamma_4 = 1/12$	$\varphi_4^{(1)} = 4\pi/3$
	$\{\mathbf{k}_4\}$	$\mathbf{k}_4^{(2)} = -\mathbf{k}_4^{(1)},$	$\gamma_4 = 1/12$	$\varphi_4^{(2)} = -4\pi/3$
	$\{\mathbf{k}_3\}$	$\mathbf{k}_3^{(3)} = -(4\mathbf{b}_1 + \mathbf{b}_2 + 2\mathbf{b}_3)/6 = \mathbf{a}_{C2/c}^* + \mathbf{b}_{C2/c}^* + \mathbf{c}_{C2/c}^* =$ $= \mathbf{a}_{C2}^* + \mathbf{b}_{C2}^* = \left\langle \frac{1}{6}\,\frac{\bar{5}}{6}\,\frac{\bar{1}}{2} \right\rangle,$	$\gamma_3 = 1/12$	$\varphi_3^{(3)} = 4\pi/3$
	$\{\mathbf{k}_3\}$	$\mathbf{k}_3^{(4)} = -\mathbf{k}_3^{(3)},$	$\gamma_3 = 1/12$	$\varphi_3^{(4)} = -4\pi/3$
	$\{\mathbf{k}_0\}$	$\mathbf{k}_0^{(4)} = (4\mathbf{b}_1 + \mathbf{b}_2 - 4\mathbf{b}_3)/12 = -\mathbf{a}_{C2/c}^* + \mathbf{b}_{C2/c}^* =$ $= -\mathbf{a}_{C2}^* + \mathbf{b}_{C2}^* + \mathbf{c}_{C2}^* = \left\langle \frac{7}{12}\,\frac{\bar{1}}{12}\,\frac{3}{4} \right\rangle,$	$\gamma_0 = \sqrt{3}/12$	$\varphi_0^{(4)} = -7\pi/6$
	$\{\mathbf{k}_0\}$	$\mathbf{k}_0^{(28)} = -\mathbf{k}_0^{(4)},$	$\gamma_0 = \sqrt{3}/12$	$\varphi_0^{(13)} = -7\pi/6$
	$\{\mathbf{k}_0\}$	$\mathbf{k}_0^{(13)} = -(8\mathbf{b}_1 + 5\mathbf{b}_2 + 4\mathbf{b}_3)/12 = \mathbf{a}_{C2/c}^* + \mathbf{b}_{C2/c}^* =$ $= \mathbf{a}_{C2}^* + \mathbf{b}_{C2}^* - \mathbf{c}_{C2}^* = \left\langle \frac{\bar{1}}{12}\,\frac{7}{12}\,\frac{3}{4} \right\rangle,$	$\gamma_0 = \sqrt{3}/12$	$\varphi_0^{(28)} = 7\pi/6$
	$\{\mathbf{k}_0\}$	$\mathbf{k}_0^{(37)} = -\mathbf{k}_0^{(13)}$	$\gamma_0 = \sqrt{3}/12$	$\varphi_0^{(37)} = 7\pi/6$

With allowance for Eq.(2) and the transition channel found, the distribution function of carbon atoms C (X) in the monoclinic (space groups $C2/c$ and $C2$) superstructures M_6C_5 (M_6X_5) depends on four long-range order parameters η_9, η_4, η_3, and η_0 corresponding to the stars $\{k_9\}$, $\{k_4\}$, $\{k_3\}$, and $\{k_0\}$, respectively, and has the form

$$
\begin{aligned}
n(x_I, y_I, z_I) = {} & y - (\eta_9/6)\cos[\pi(x_I - y_I + z_I)] - \\
& - (\eta_4/12)\{\cos[4\pi(x_I + y_I)/3] - (\sqrt{3})\sin[4\pi(x_I + y_I)/3]\} - \\
& - (\eta_3/12)\{\cos[\pi(x_I - 5y_I - 3z_I)/3] - (\sqrt{3})\sin[\pi(x_I - 5y_I - 3z_I)/3]\} - \\
& - (\eta_0/12)\{3\cos[\pi(x_I + 7y_I + 9z_I)/6] - (\sqrt{3})\sin[\pi(x_I + 7y_I + 9z_I)/6] + \\
& + 3\cos[\pi(7x_I + y_I - 9z_I)/6] - (\sqrt{3})\sin[\pi(7x_I + y_I - 9z_I)/6]\}.
\end{aligned}
\tag{4}
$$

Distribution function (4) describing the monoclinic superstructures M_6C_5 assumes five different values n_1, n_2, n_3, n_4, and n_5 at all sites of the basic nonmetallic fcc sublattice (Table 3). The same distribution function means that only one of the two monoclinic (space groups $C2/c$ and $C2$) models of the structure is valid for the M_6C_5 phase. Since the unit cells have the same volume, the unit cell with a higher symmetry should be chosen among the two monoclinic unit cells. The point symmetry group $2/m$ (C_{2h}) of the monoclinic (space group $C2/c$) M_6C_5 phase includes four symmetry elements h_1, h_4, h_{25}, and h_{28}, while the point group 2 (C_2) of the monoclinic (space group $C2$) M_6C_5 superstructure includes only two symmetry elements h_1 and h_4. Therefore, the monoclinic (space group $C2/c$) model of the structure of the M_6C_5 phase is characterized by a higher symmetry as compared to the monoclinic model of the M_6C_5 phase with space group $C2$. Thus, the monoclinic (space group $C2/c$) model describes the crystal structure of ordered phase M_6C_5 ($M_6C_5\square$) more correctly than the model with space group $C2$. This means that the monoclinic (space group $C2$) M_6C_5 superstructures experimentally determined in [15, 16, 18, 19] and described later in reviews [1, 4, 40, 41] actually belong to space group $C2/c$.

4.2. Monoclinic (Space Group $C2/m$) Superstructure M_6C_5

The monoclinic (space group $C2/m$) unit cells of the M_6C_5 superstructure shown in Figure 3 involves two formula units M_6C_5. The translation vectors and the coordinates of atoms and vacancies in the perfect monoclinic (space group $C2/m$) M_6C_5 superstructure are given in Table 5.

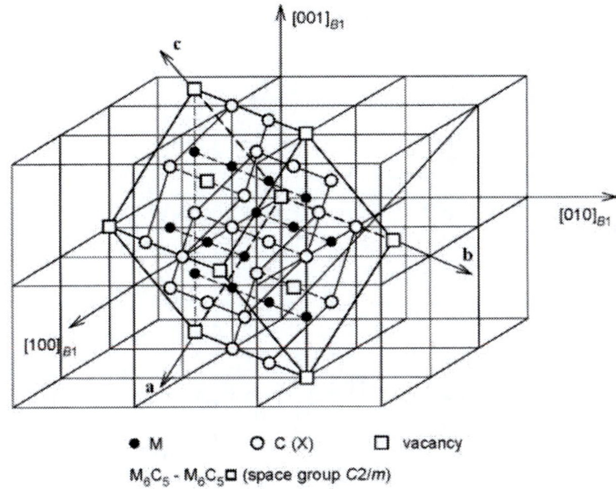

Figure 3. Monoclinic (space group $C2/m$) unit cell of the M$_6$C$_5$ superstructure in a lattice with the B1 structure: ($\bullet$) metal atoms M, ($\circ$) interstitial carbon atoms C, and ($\square$) vacancies.

The basis vectors of the reciprocal lattice of the monoclinic (space group $C2/m$) M$_6$C$_5$ superstructure are $\mathbf{a}^*_{C2/m} = \frac{1}{2}\langle 1\,\overline{1}\,\overline{1}\rangle$, $\mathbf{b}^*_{C2/m} = \frac{1}{3}\langle 110\rangle$, and $\mathbf{c}^*_{C2/m} = \frac{1}{2}\langle 1\,\overline{1}\,1\rangle$. The point symmetry group $2/m$ (C_{2h}) of the monoclinic (space group $C2/m$) M$_6$C$_5$ phase includes four symmetry elements h_1, h_4, h_{25}, and h_{28}. The disorder-order transition channel associated with the formation of the monoclinic (space group $C2/m$) superstructure M$_6$C$_5$ involves five nonequivalent superstructure vectors of three stars $\{\mathbf{k}_9\}$, $\{\mathbf{k}_4\}$ and $\{\mathbf{k}_3\}$ (Table 6). Therefore, the distribution function of carbon atoms in the monoclinic (space group $C2/m$) superstructure M$_6$C$_5$ depends on three long-range order parameters η_9, η_4, and η_3:

$$n(x_I, y_I, z_I) = y - (\eta_9/6)\cos[\pi(x_I - y_I + z_I)] - (\eta_4/3)\cos[4\pi(x_I + y_I)/3] -$$
$$- (\eta_3/3)\cos[\pi(x_I - 5y_I - 3z_I)/3] \,. \tag{5}$$

Like distribution function (2) describing the trigonal superstructure M$_6$C$_5$ (see Table 1) the distribution function (5) takes on the same four values at all sites of the basic nonmetallic fcc sublattice (Table 5).

Table 5. Monoclinic M_6C_5-type superstructures (space group No. 12: $C2/m$ ($C12/m1$ (C_{2h}^3))):

$$Z = 2, \mathbf{a}_{C2/m} = \tfrac{1}{2}\langle 1\,\bar{1}\,2 \rangle_{B1}, \quad \mathbf{b}_{C2/m} = \tfrac{1}{2}\langle 3\,3\,0 \rangle_{B1}, \quad \mathbf{c}_{C2/m} = \tfrac{1}{2}\langle 1\,\bar{1}\,2 \rangle_{B1}$$

Atom	Position and multiplicity	Atomic coordinates in the ideal ordered structure			Atomic coordinates in the basic disordered $B1$ structure (according to Figure 3)			Probabilities of detecting C atom at different positions of M_6C_5 superstructure, i. e., values of the distribution function $n(x_1, y_1, z_1)$ (5) for C atoms
		$x/a_{C2/m}$	$y/a_{C2/m}$	$z/a_{C2/m}$	x/a_{B1}	y/a_{B1}	z/a_{B1}	
C1 (vacancy)	Special 2(a)	0	0	0	0	0	0	$n_1 = y - \eta_9/6 - \eta_4/3 - \eta_3/3$
C2	Special 2(d)	0	1/2	1/2	1	1/2	1/2	$n_2 = y + \eta_9/6 - \eta_4/3 + \eta_3/3$
C3	Special 4(g)	0	1/3	0	½	1/2	0	$n_3 = y - \eta_9/6 + \eta_4/6 + \eta_3/6$
C4	Special 4(h)	0	1/6	1/2	½	0	1/2	$n_4 = y + \eta_9/6 + \eta_4/6 - \eta_3/6$
M1	Special 4(i)	1/4	0	3/4	½	-1/2	1/2	
M2	General 8(j)	1/4	2/3	3/4	3/2	1/2	1/2	

Table 6. $MC_y - M_6C_5$ disorder-order phase transition channel and parameters of the distribution function $n(x_I, y_I, z_I)$ (5) describing the monoclinic (space group $C2/m$) M_6C_5 superstructure

Space group	Disorder-order phase transition channel		Parameters of the distribution function (5)	
	wave-vector star $\{\mathbf{k}_s\}$	rays $\mathbf{k}_s^{(j)}$ of the star $\{\mathbf{k}_s\}$	γ_s	$\varphi_s^{(j)}$
$C2/m$	$\{\mathbf{k}_9\}$	$\mathbf{k}_9^{(3)} = \mathbf{b}_2/2 = \mathbf{c}_{C2/m}^* = \left\langle \frac{1}{2}\, \frac{\bar{1}}{2}\, \frac{1}{2} \right\rangle,$	$\gamma_9 = 1/6$	$\varphi_9^{(3)} = \pi$
	$\{\mathbf{k}_4\}$	$\mathbf{k}_4^{(1)} = (\mathbf{b}_1 + \mathbf{b}_2 + 2\mathbf{b}_3)/3 = \mathbf{b}_{C2/m}^* = \left\langle \frac{2}{3}\, \frac{2}{3}\, 0 \right\rangle,$	$\gamma_4 = 1/6$	$\varphi_4^{(1)} = \pi$
	$\{\mathbf{k}_4\}$	$\mathbf{k}_4^{(2)} = -\mathbf{k}_4^{(1)}$	$\gamma_4 = 1/6$	$\varphi_4^{(2)} = -\pi$
	$\{\mathbf{k}_3\}$	$\mathbf{k}_3^{(3)} = -(4\mathbf{b}_1 + \mathbf{b}_2 + 2\mathbf{b}_3)/6 = \mathbf{a}_{C2/m}^* - \mathbf{b}_{C2/m}^* = \left\langle \frac{1}{6}\, \frac{\bar{5}}{6}\, \frac{\bar{1}}{2} \right\rangle,$	$\gamma_3 = 1/6$	$\varphi_3^{(3)} = \pi$
	$\{\mathbf{k}_3\}$	$\mathbf{k}_3^{(4)} = -\mathbf{k}_3^{(3)}$	$\gamma_3 = 1/6$	$\varphi_3^{(4)} = -\pi$

However, the relative arrangement of sites of four sublattices in the monoclinic (space group $C2/m$) ordered structure differs from that in the trigonal (space group $P3_112$) superstructure M_6C_5. The long-range order in the distribution of carbon atoms and vacancies in the examined monoclinic (space group $C2/m$) superstructure M_6C_5 also differs from that in the monoclinic (space group $C2/c$) superstructure M_6C_5.

5. SEQUENCE OF M_6C_5 SUPERSTRUCTURES FORMATION

The disorder-order or order-order transformations occurring with a decrease in the temperature are transitions from a state with a higher free energy to a state with a lower free energy. The state of matter in atomic or atomic-vacancy ordering can be characterized by the Landau thermodynamic potential, which in this case is a functional of the probabilities of finding atoms of a particular type at lattice sites, the site coordinates, and the temperature. In turn, the probabilities are functions of the long-range order parameters. The Landau potential exhibits several minima corresponding to the high-symmetry disordered and the low-symmetry ordered phases. With a decrease in temperature, the transition from the disordered phase to any of the ordered phases or from one ordered phase to another ordered phase is accompanied by a reduction of symmetry. Symmetry analysis makes it possible to quantitatively determine the reduction of symmetry when a particular superstructure is formed and to reveal the physically admissible sequence of the formation of these superstructures.

The available literature data and the performed symmetry analysis of M_6C_5 superstructures show that the trigonal (space group $P3_112$ (No. 155)) and two monoclinic (space group $C2/m$ (No. 12) and space group no. $C2/c$ (No. 15)) M_6C_5 phases form as a result of ordering of nonstoichiometric carbides MC_y with $y \approx 5/6$. Ordering of C atoms and structural vacancies □ occurs in the basic nonmetallic fcc sublattice of the disordered cubic (space group $Fm\bar{3}m$) phase MC_y and is associated with splitting of the $4(b)$ high-symmetry positions into two or more positions of the low-symmetry ordered phase. The $4(b)$ position has a symmetry point group $m\bar{3}m$ (O_h) which includes 48 symmetry elements h_1-h_{48} [1, 2, 44, 45].

The symmetry point groups of the trigonal (space group $P3_112$), monoclinic (space group $C2/c$), and monoclinic (space group $C2/m$) superstructures M_6C_5 include six (h_1, h_5, h_9, h_{13}, h_{17}, h_{21}), four (h_1, h_4, h_{25}, h_{28}), and four (h_1, h_4, h_{25}, h_{28}) symmetry elements, respectively, and are the subgroups of the point group of the basic disordered cubic (space group $Fm\bar{3}m$) phase MC_y. Therefore, a transition from the disordered carbide to any of these phases is a disorder-order phase transformation. The three M_6C_5-type superstructures under consideration are formed with distortion of symmetry according to three or four irreducible representations. This means that the phase transitions $MC_y \rightarrow M_6C_5$ do not satisfy the Landau theory-group criterion for second-order phase transitions and occur via the first-order transition mechanism.

The disorder-order or order-order transformations in nonstoichiometric compounds occur with a reduction of the point symmetry of the crystal. Indeed, some symmetry transformations of the high-symmetry phase that bring mutually substituted atoms of the solid solution (or the occupied and vacant sites of the nonstoichiometric compound) into coincidence are not involved in the group of symmetry elements of the low-symmetry ordered crystal, because these sites become crystallographically nonequivalent.

The symmetry point group 322 (D_3) of the trigonal (space group $P3_112$) superstructure M_6C_5 includes six symmetry elements h_1, h_5, h_9, h_{13}, h_{17} and h_{21}, while the point group $m\bar{3}m$ (O_h) of the basic disordered cubic phase MC_y involves 48 elements h_1-h_{48} [1, 2, 44, 45]. Therefore, the rotational reduction of symmetry is equal to 8. The reduction of translational symmetry is equal to the ratio between the volumes of the unit cells of the low-symmetry and high-symmetry phases and amounts to 4.5 upon transition from the high-symmetry disordered carbide MC_y to the low-symmetry trigonal carbide M_6C_5. The total reduction of symmetry is the product of the rotational and translational reductions of symmetry. Therefore, in the transition from the disordered nonstoichiometric cubic (space group $Fm\bar{3}m$) carbide MC_y to the ordered trigonal (space group $P3_112$) phase M_6C_5, the total reduction of symmetry is $N = n(G)/n(G_D) = 36$ where $n(G)$ and $n(G_D)$ are the orders of the space group G of the high-symmetry phase and the space group G_D of the low-symmetry phase, respectively.

The point symmetry group $2/m$ (C_{2h}) of the monoclinic (space groups $C2/m$ and $C2/c$) ordered phases M_6C_5 includes four symmetry elements h_1, h_4, h_{25}, and h_{28}. Therefore, the formation of both monoclinic M_6C_5 superstructures is accompanied by a 12-fold decrease in their rotational symmetry. The reduction of translational symmetry upon transition from the disordered carbide MC_y to the monoclinic (space group $C2/m$) phase M_6C_5 is 3 and to the monoclinic (space group $C2/c$) phase M_6C_5 is 6. The total reduction of symmetry upon transitions "disordered carbide MC_y (space group $Fm\bar{3}m$) → ordered monoclinic (space group $C2/m$) phase M_6C_5" and "disordered carbide MC_y (space group $Fm\bar{3}m$) → ordered monoclinic (space group $C2/c$) phase M_6C_5" is 36 and 72, respectively.

As regards the transitions between particular superstructures M_6C_5, it is clear from the relationships between elements h_i that the trigonal superstructure is not symmetry-related to the monoclinic superstructures M_6C_5, because its point group is not a group or a subgroup of the point groups of the monoclinic superstructures. Consequently, the transition between the trigonal and either of the two monoclinic phases M_6C_5 cannot be an order-order transformation, but can occur as a polymorphic transformation.

The order-order transformation is possible only for the monoclinic (space groups $C2/m$ and $C2/c$) M_6C_5 superstructures. Since the point symmetry groups of these superstructures coincide with each other, the rotational symmetry does not change as a result of transition from one symmetry group to the other. The reduction of translational symmetry upon order-order transition is equal to the ratio between the unit cell volumes or the number of the sites in the unit cells of the low-symmetry and high-symmetry ordered phases. For the examined monoclinic (space groups $C2/m$ and $C2/c$) superstructures M_6C_5, the decrease in the translational symmetry equal to 2 will occur as a result of order-order transition of the monoclinic (space group $C2/m$) phase M_6C_5 to the monoclinic (space group $C2/c$) phase M_6C_5. Therefore, the total reduction of symmetry upon the transitions "ordered monoclinic (space group $C2/m$) phase M_6C_5 → ordered monoclinic (space group $C2/c$) phase M_6C_5" is equal 2.

Thus, a decrease in temperature can lead to two sequences of transformations associated with the phases M_6C_5 (Figure 4). The first sequence "cubic (space group $Fm\bar{3}m$) phase MC_y → monoclinic (space group $C2/m$) phase M_6C_5 → monoclinic (space group $C2/c$) phase M_6C_5" consists of transformations of the disordered cubic phase MC_y to the ordered monoclinic (space group $C2/m$) phase M_6C_5 and then to the ordered monoclinic (space group $C2/c$) phase M_6C_5 and involves only disorder-order and order-order transformations. The alternative sequence includes the disorder-order transformation from the disordered cubic (space group $Fm\bar{3}m$) phase MC_y to the ordered trigonal (space group $P3_112$) phase M_6C_5 and the polymorphic transformation of the trigonal ordered phase M_6C_5 to the monoclinic (space group $C2/c$) phase M_6C_5.

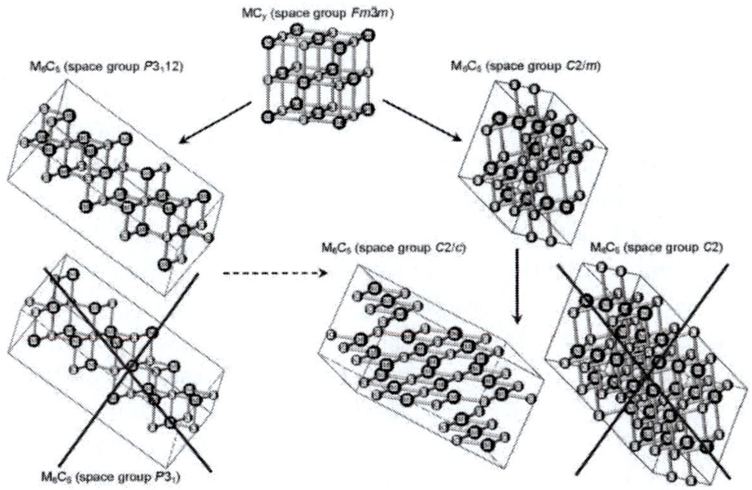

Figure 4. The physically admissible sequences of the disorder-order and order-order phase transformations associated with the formation of the M_6C_5-type superstructures in strongly nonstoichiometric carbides MC_y with the $B1$ structure. The unit cell of a nonstoichiometric carbide MC_y with the basic cubic (space group $Fm\bar{3}m$) structure is shown at the top, nonmetallic sublattice sites of this unit cell are statistically occupied by C atoms with a probability y. ($\bullet$) metal atoms M, ($\bigcirc$) interstitial carbon atoms C, vacant sites of the nonmetallic sublattices of the different M_6C_5 superstructures are not shown.

If one takes into account only the basic nonmetallic fcc sublattice, a common feature of the M_6C_5 superstructures under discussion is that atomic planes $(1\bar{1}1)_{fcc}$ of two types alternate in the $[1\bar{1}1]_{fcc}$ direction. Using the site occupation probabilities found (see Tables 1, 3, and 5), one can readily determine the probabilities of detecting a carbon atom at the site of these nonmetallic planes. For all superstructures M_6C_5, the probability of detecting a carbon atom at the site of the plane of the first type is $P_1 = y - \eta_9/6$, that at the site of the plane of the second type $P_2 = y + \eta_9/6$. Thus, the carbon-atom occupancy of the sites of the $(1\bar{1}1)_{fcc}$ nonmetallic planes depends only on the long-range order parameter η_9. It is clear that when $\eta_9 = 1$ the defective and the complete carbon $(1\bar{1}1)_{fcc}$ planes alternate; in the defective planes, one third of sites is vacant and each vacancy $\square$ is surrounded by six C atoms. The superstructure vector $\mathbf{k}_9^{(3)}$ is responsible for the alteration of the $(1\bar{1}1)_{fcc}$

nonmetallic planes of two types. Others superstructure vectors are responsible for the direction of relative displacement of carbon atoms and vacancies of the equidistant defective $(1\bar{1}1)_{\text{fcc}}$ nonmetallic planes of the examined M_6C_5 superstructures.

The effect of the disorder-order and order-order phase transformations on the properties of nonstoichiometric carbides is appreciable enough [51]. Such transformations are used for the formation of a nanostructure in compact and dispersed strongly nonstoichiometric carbides [52-54]. The formation of a nanostructure upon ordering is due to partition of the basic disordered phase grains to the domains of the ordered phase. For example, in nonstoichiometric vanadium carbide, nanostructures with the average size of the ordered phase domains to 120, 60, and 20 nm are formed as a result of ordering under different heat treatment conditions [55, 56]. Apparently, additional decrease in the domain sizes, i.e. an even finer nanostructure, can be attained by ensuring the conditions for the order-order transformations.

6. SHORT-RANGE ORDER IN M_6C_5 SUPERSTRUCTURES

Short-range order describes radial distribution of atoms in the lattice of a solid solution (or of interstitial atoms and vacancies in case of nonstoichiometric compounds) and determines the conditional probability that an atom of a certain species is present at a given site. Long-range order includes both radial and angular distributions and allows determining the real probability that a site in a sublattice is filled with an atom of one species or another. Although short- and long-range orders differ fundamentally, there is a certain interrelation between them. If a solid solution exhibits long-range order, it certainly has short-range order, too. In this case short-range order parameters α_j for any j-th coordination sphere are constant quantities, which assume a fixed values for a completely ordered structure (for a completely ordered solid solution, $\alpha_j = \alpha_j^{\text{ord}}$ is constant). Short-range order, which accompanies long-range order, is called a superstructural short-range order.

For the nonstoichiometric compound MX_y ($MX_y\square_{1-y}$) or the solid solution A_yB_{1-y}, short-range order in the j-th coordination sphere of nonmetallic sublattice is characterized by the parameter α_j, which has the form [1]:

$$\alpha_i = 1 - P_{\square X}^{(j)} / P_{\square X}^{\text{bin}} \equiv 1 - P_{BA}^{(j)} / P_{BA}^{\text{bin}} , \qquad (6)$$

where $P_{\square X}^{\text{bin}} = y(1 - y)$ is the binomial probability of formation of "vacancy - interstitial atom" $\square X$ pair, which corresponds to a disordered state and depends only on the composition of the nonstoichiometric compound MX_y; $P_{\square X}^{(j)}$ is the real probability of formation of "vacancy - interstitial atom" $\square X$ pairs, in which the distance between vacant site $\square$ and atom X is equal to the radius of the j-th coordination sphere. The vacancy $\square$ is the center of the coordination spheres.

The presence of the distribution functions $n(\mathbf{r})$ of carbon atoms in M_6C_5 superstructures allows one to determine the superstructural short-range order parameters in any coordination sphere quantitatively. Let an ordering nonstoichiometric compound MX_y have a superstructure where a unit cell contains M types of nonequivalent positions with multiplicities g_f ($\displaystyle\sum_{f=1}^{M} g_f = \Phi$

being the number of sites of a nonmetallic sublattice in the superstructure unit cell). According to Rempel and Gusev [57, 58], the superstructural short-range order parameter $\alpha_j\{\eta\}$ depends on the long-range order parameters η and is equal to

$$\alpha_j\{\eta\} = 1 - \frac{\displaystyle\sum_{f=1}^{M} g_f[1 - n(\mathbf{r}_f)]\sum_{i=1}^{z_j} n(\mathbf{r}_{if}^{(j)})}{y(1-y)z_j \displaystyle\sum_{f=1}^{M} g_f} , \qquad (7)$$

where $n(\mathbf{r}_{if}^{(j)})$ is the distribution function value for species X atoms at sites $\mathbf{r}_{if}^{(j)}$ belonging to the j-th coordination sphere of nonmetallic sublattice, and z_j is the coordination number of the j-th coordination sphere. If the distribution function $n(\mathbf{r})$ is known, it is easy to find from Eq.(7) the short-range order parameters $\alpha_j\{\eta\}$ in a nonstoichiometric compound of an arbitrary composition for arbitrarily remote coordination spheres when long-range order parameters change from zero to a maximum value $\eta_{\max}$. In fact, determination of the parameter $\alpha_j\{\eta\}$ from Eq.(7) reduces to computation of the average number of atoms of a particular species over a crystal volume in the j-th coordination sphere of an atom of the other species. In this respect relationship (7) is identical to the Kauli formula [59].

Using distribution functions (3)-(5) and coordinates of sites **r** forming a particular coordination sphere, the superstructural short-range order parameters are calculated from Eq.(7) for the coordination spheres of completely ordered stoichiometric superstructures, i.e. with all long-range order parameters being unity and $y = y_{st} = (2t - 1)/2t = 5/6$. The calculated $\alpha_j(y_{st}, \eta_{max})$ values are given in Table 7.

Table 7. Short-range order parameters $\alpha_j(y_{st}, \eta_{max})$ for the trigonal (space group $P3_112$) and monoclinic (space groups $C2/c$ and $C2/m$) superstructures M_6C_5 with a basic nonmetallic fcc sublattice

[1] Coordination sphere			$\alpha_j(y_{st}, \eta_{max})$
j	hkl	[2] z_j	
1	110	12	$-1/5$
2	200	6	$-1/5$
3	211	24	$1/5$
4	220	12	0
5	310	24	0
6	222	8	$-1/5$
7	321	48	$-1/10$
8	400	6	$1/5$
9	330, 411	$12 + 24 = 36$	$2/15$
10	420	24	$-1/5$
11	332	24	0
12	422	24	$2/5$
13	431, 510	$48 + 24 = 72$	$-1/15$
14	521	48	0
15	440	12	0

[1] The vacancy □ is the center of the coordination spheres; the radius of the j-th coordination sphere is $R_j = (a_{B1}/2)\sqrt{h^2 + k^2 + l^2}$ where a_{B1} is the basic lattice constant.

[2] z_j is the coordination number.

In accordance with our calculations, the short-range order parameters $\alpha_j(y_{st}, \eta_{max})$ for the three M_6C_5 superstructures with different crystal lattices coincide completely. Identical short-range order in superstructures of different symmetry suggests that in addition to strong central interatomic interactions, noncentral long-range forces leading to different symmetries for ordered M_6C_5 phases should be significant in nonstoichiometric compounds MC_y (MX_y). Thus, even the maximum short-range order seems to be insufficient for the formation of such ordered structures. This is in agreement with the findings [60, 61].

CONCLUSION

The analysis performed showed that the identity or difference in the proposed superstructure models can be revealed only by determining the disorder-order phase transition channel. If different models have the same transition channel, only that superstructure model is valid, which features the highest point symmetry within the considered crystalline system.

The performed symmetry analysis of M_6C_5 superstructures and critical review of literature data on the structure of ordered nonstoichiometric vanadium, niobium, and tantalum carbides give us grounds to believe that a decrease in temperature can lead to two sequences of transformations associated with the formation of M_6C_5 phases. The first sequence, i.e., the transformation from the disordered cubic (space group $Fm\bar{3}m$) phase MC_y to the ordered monoclinic (space group $C2/m$) phase M_6C_5 and then to the ordered monoclinic (space group $C2/c$) phase M_6C_5, involves only the disorder-order and order-order transformations which occur with a reduction of the symmetry by factors 36 and 2, respectively. An alternative sequence, namely, the transformation from the disordered cubic (space group $Fm\bar{3}m$) phase MC_y to the ordered trigonal (space group $P3_112$) phase M_6C_5 and then to the monoclinic (space group $C2/c$) phase M_6C_5, includes the disorder-order transformation, which occurs with reduction of the symmetry by a factor of 36, and the polymorphic transformation of the trigonal phase to the monoclinic phase. The first sequence is more probable as judged from the experimental data. It can be assumed that the factors responsible for the occurrence of a particular sequence are related to the macroscopic state of nonstoichiometric carbides, namely, the size and morphology of grains of the disordered phase and the onset of the formation of the primary ordered phase on a specific crystallographic surface.

Completely ordered monoclinic (space groups $C2/m$ and $C2/c$) and trigonal (space group $P3_112$) superstructures M_6C_5 have the same short-range order in nonmetal sublattice.

ACKNOWLEDGMENTS

The authors are grateful to the Russian Foundation for Basic Research (Grant No. 13-03-00077a) for financial support.

REFERENCES

[1] Gusev, A.I; Rempel, A.A.; Magerl, A.J. *Disorder and Order in Strongly Nonstoichiometric Compounds: Transition Metal Carbides, Nitrides and Oxides*. Springer: Berlin – Heidelberg – New York – London, 2001. 607 pp.

[2] Gusev, A.I. *Nonstoichiometry, Disorder, Short-Range and Long-Range in Solids*. Nauka-Fizmatlit: Moscow, 2007. 856 pp (*in Russian*)

[3] Rogl, P. Pd-B-C. 1998 In: Effenberg, G.; Rogl, P.; Ilenko, S. *Phase diagrams of ternary metal-boron-carbon systems*. ASM International: Materials Park, OH, 1998, pp 234-241.

[4] Gusev, A.I.; Rempel A.A. Superstructures of non-stoichiometric interstitial compounds and the distribution functions of interstitial atoms. *Physica status solidi* (*a*), 1993, 135, No 1, 15-58.

[5] Gusev, A.I. Sequence of phase transformations in the formation of superstructures of the M_6C_5 type in nonstoichiometric carbides. *Zh. Eksp. Teor. Fiz.*, 2009, 136, No 3, 486-504 (*in Russian*). (English Transl: *JETP*, 2009, 109, No 3, 417-433)

[6] Venables, J.D.; Kahn, D.; Lye, R.G. Structure of ordered compound V_6C_5. *Philosoph. Mag.*, 1968, 18, No 151, 177-192.

[7] Venables, J.D.; Lye, R.G. Radiation damage of ordered V_6C_5 by electron microscope beam bombardment. *Philosoph. Mag.*, 1969, 19, No 159, 565-582.

[8] Kahn, D.; Lye, R.G. Nuclear magnetic resonance rotation patterns in single-crystal V_6C_5. *Bull. Amer. Phys. Soc.*, 1969, 14, No 3, 332-333.

[9] Karimov, I,; Faizullaev, F.; Kalanov, M.; Emiraliev, A.; Rakhimov, A.S.; Slepoi, L.; Polischuk, V.S. Neutron diffraction study of trigonal ordering of $VC_{0.814}$ vanadium carbide. *Doklady AN Uzb. SSR*, 1976, No 2, 32-33 (*in Russian*).

[10] Karimov, I.; Faizullaev, F.; Kalanov, M.; Emiraliev, A.; Polischuk, V.S. Phase transformations in vanadium monocarbides. *Izv. AN Uzb. SSR. Seriya Fiz.-Mat. Nauk*, 1978, No 4, 87-88 (*in Russian*).

[11] Lipatnikov, V.N.; Lengauer, W.; Ettmayer, P.; Keil, E.; Groboth, G.; Kny, E. Effects of vacancy ordering on structure and properties of vanadium carbide. *J. Alloys Comp.*, 1997, 261, 192-197.

[12] Khaenko, B.V.; Kukol, V.V. Structural characteristics of V_6C_5 carbide. *Doklady AN Ukr. SSR. Seriya A. Fiz.-Mat. I Tekhn. Nauki*, 1987, No 1, 78-82 (*in Russian*).

[13] Khaenko, B.V.; Kukol, V.V.; Ershova, L.S. Ordering of vanadium carbide. *Izv. AN SSSR. Neorgan. Materialy*, 1989, 25, No 2, 263-269 (*in Russian*).

[14] Cenzual, K.; Gelato, L.V.; Penzo, M.; Parthé, E. Inorganic structure types with revised space groups. I. *Acta Crystallogr. B*, 1991, 47, No 4, 433-439.

[15] Billingham, J.; Bell, P.S.; Lewis, M.H. Superlattice with monoclinic symmetry based on compounds V_6C_5. *Philosoph. Mag.*, 1972, 25, No 3, 661-671.

[16] Billingham, J.; Bell, P.S.; Lewis, M.H. Vacancy short-range order in substoichiometric transition metal carbides and nitrides with the NaCl structure. Electron diffraction studies of short-range ordered compounds. *Acta Crystallogr. A*, 1972, 28, No 6, 602-606.

[17] Lewis, M.H.; Billingham, J.; Bell, P.S. Non-stoichiometry in ceramic compounds. In: *Solid State Chemistry, Proc. of 5th Intern. Mater. Res. Symp. (NBS Special Publ. 364)*. NBS Publ.: Berkley, CA, 1972, pp 1084-1114.

[18] Hiraga, K. Vacancy ordering in vanadium carbide based on V_6C_5. *Philosoph. Mag.*, 1973, 27, No 6, 1301-1312.

[19] Karimov, I.; Faizullaev, F.; Kalanov, M.; Emiraliev, A.; Polischuk, V.S. Monoclinic ordering of $VC_{0.839}$ vanadium monocarbide. *Izv. AN Uzb. SSR. Seriya Fiz.-Mat. Nauk*, 1976, No 4, 74-77 (*in Russian*).

[20] Kesri, R.; Hamar-Thibault, S. Structures ordonnees a longue distance dans les carbures MC dans les fontes. *Acta Met.* 1988, 36, No 1. 149-166.

[21] Adnane, L.; Kesri, R.; Hamar-Thibault, S. Vanadium carbides formed from the melt by solidification in Fe-V-X-C alloys (X = Cr, Mo, Nb). *J. Alloys Comp.*, 1992, 178, No 1-2, 71-84.

[22] Epicier, T.; Acevedo, D.; Perez, M. Crystallographic structure of vanadium carbide precipitates in a model Fe-C-V steel. *Philosoph. Mag.*, 2008, 88, No 1, 31-45.

[23] Rempel, A.A.; Gusev, A.I.; Zubkov, V.G.; Shveikin, G.P. Structure of ordered niobium carbide Nb_6C_5. *Doklady AN SSSR*, 1984, 275,

No 4, 883-887 (*in Russian*). (English Transl: *Sov. Phys. Dokl.*, 1984, 29, No 4, 257-259).

[24] Gusev, A.I.; Rempel, A.A. Ordering in the carbon sublattice of nonstoichiometric niobium carbide. *Fiz. Tverd. Tela*, 1984, 26, No 12, 3622-3627 (*in Russian*). (Engl. Transl.: *Sov. Physics - Solid State*, 1984, 26, No 12, 2178-2181).

[25] Rempel, A.A.; Gusev, A.I.Order-disorder phase transition in nonstoichiometric niobium carbide. *Kristallografiya*, 1985, 30, No 6, 1112-1115 (*in Russian*). (Engl. Transl.: *Sov. Physics - Crystallography*, 1985, 30, No 6, 648-650).

[26] Lipatnikov, V.N.; Gusev, A.I.; Ettmayer, P.; Lengauer, W. Phase transformation in non-stoichiometric vanadium carbide. *J. Phys.: Condens. Matter*, 1999, 11, No 1, 163-184.

[27] Lipatnikov, V.N.; Gusev, A.I.; Ettmayer, P.; Lengauer, W. Order-disorder phase transformations and specific heat of nonstoichiometric vanadium carbide. *Fiz. Tverd. Tela*, 1999, 41, No 3, 529-536 (*in Russian*). (Engl. Transl.: *Physics of the Solid State*, 1999, 41, No 3, 474-480).

[28] Venables, J.D.; Meyerhoff, M.H. Ordering effect in NbC and TaC. 1972 In: *Solid State Chemistry. Proc. of 5th Intern. Mater. Res. Symp.* (*NBS Special Publ. 364*) NBS Publ.: Berkley, CA, 1972, pp 583-590.

[29] Landesman, J.P.; Christensen, A.N.; de Novion, C.H.; Lorenzelli, N.; Convert, P. Order-disorder transition and structure of the ordered vacancy compound Nb_6C_5: powder neutron diffraction studies. *J. Phys. C: Solid State Phys.*, 1985, 18, No 4, 809-823.

[30] Christensen, A.N. Vacancy order in Nb_6C_5. *Acta chem. scand.* A, 1985, 39, No 10, 803-804.

[31] Gusev, A.I.; Rempel, A.A. Order-disorder phase transition channel in niobium carbide. *Physica status solidi* (a), 1986, 93, No 1, 71-80.

[32] Rempel, A.A.; Gusev, A.I.; Belyaev, M.Yu. ^{93}Nb NMR study of an ordered and a disordered nonstoichiometric niobium carbide. *J. Phys. C: Solid State Physics*, 1987, 20, No 34, 5655-5666.

[33] Arbuzov, M.P.; Khaenko, B.V.; Sivak, O.P. Superstructures of order in niobium monocarbide. *Doklady AN UkrSSR. Seriya A*, 1984, No 10, 86-88 (*in Russian*).

[34] Khaenko, B.V.; Sivak, O.P.; Sinelnikova, V.S. X-ray study of ordered modification of niobium monocarbide. *Izv. AN SSSR. Neorgan. Materialy*, 1984, 20, No 11, 1825-1828 (*in Russian*).

[35] Khaenko, B.V.; Sivak, O.P. Strusture of order of niobium monocarbide. *Kristallografiya*, 1990, 35, No 5, 1110-1115 (*in Russian*).

[36] Gusev, A.I.; Rempel, A.A.; Lipatnikov, V.N. Incommensurate ordered phase in nonstoichiometric tantalum carbide. *J. Phys.: Condens. Matter*, 1996, 8, No 43, 8277-8293.

[37] Rempel, A.A.; Lipatnikov, V.N.; Gusev, A.I. The superstructure in nonstoichiometric tantalum carbide. *Doklady AN SSSR*, 1990, 310, No 4, 878-882 (*in Russian*). (Engl. Transl.: *Sov. Physics Doklady*, 1990, 35, No 2, 103-106).

[38] Lipatnikov,V.N.; Rempel, A.A. Formation of the incommensurate ordered phase in TaC_y carbide. *Pis'ma v ZhETF*, 2005, 81, No 7, 410-414 (*in Russian*). (English Transl: 2005 *JETP Letters*, 2005, 81, No 7, 326-330).

[39] Gusev, A.I.; Rempel, A.A. Phase diagrams of metal-carbon and metal-nitrogen systems and ordering in strongly nonstoichiometric carbides and nitrides. *Physica status solidi* (a), 1997, 163, No 2, 273-304.

[40] Gusev, A.I.; Rempel, A.A. Atomic ordering and phase equilibria in strongly nonstoichiometric carbides and nitrides. In: Gogotsi Y.G., Andrievski R.A., editors. *Materials Science of Carbides, Nitrides and Borides*. Kluwer Academic Publishers: Netherlands, 1999, pp 47-64.

[41] Gusev, A.I. Order-disorder transformations and phase equilibria in strongly nonstoichiometric compounds. *Usp.Fiz.Nauk*, 2000, 170, No 1, 3-40 (*in Russian*). (English Transl: *Physics - Uspekhi*, 2000, 43, No 1, 1-37).

[42] Gusev, A.I. Long- and short-range order in the Pd_6B monoclinic superstructure and M_6X_5 and M_6X allied superstructures. *Zh. Eksp. Teor. Fiz.*, 2011, 140, No 1 (7), 112-122 (*in Russian*). (English Transl: 2011 *JETP*, 2011, 112, No 7, 96-105).

[43] Gusev, A.I. Symmetry analysis of the monoclinic Pd_6B superstructure: long- and short-range orders. *Fiz. Tverd. Tela*, 2011, 53, No 8, 1582-1588 (*in Russian*). (English Transl: *Phys. Solid State*, 2011, 53, No 8, 1664-1671).

[44] Kovalev, O.V. *Irreducible Representations of the Space Groups*. Gordon and Breach: New York, 1965. 210 pp.

[45] Kovalev, O.V. *Representations of the Crystallographic Space Groups: Irreducible Representations, Induced Representations and Corepresentation*. 2nd edition. Gordon and Breach Science Publi.: Yverdon - Paris - Berlin - London - Tokyo – Amsterdam, 1993. 390 pp.

[46] Khachaturyan, A.G. *Theory of Structural Transformations in Solids*. John Wiley and Sons: New York, 1983. 574 pp.

[47] Vainshtein, B.K. *Modern Crystallography*, Vol. 1: *Fundamentals of Crystals: Symmetry and Methods of Structural Crystallography*. Nauka: Moscow, 1979, pp 182-222 (*in Russian*).

[48] Berger, T.G. *Phase Transformations in Interstitial* Pd-B *Alloys* (Dissertation an der Universität Stuttgart). Institut für Metallkunde der Universität Stuttgart und Max-Planck-Institut für Metallforschung: Stuttgart, 2005. 107 pp.

[49] Berger, T.G.; Leineweber, A.; Mittemeijer, E.J.; Sarbu, C.; Duppel, V.; Fischer, P. On the formation and crystal structure of the Pd_6B phase. *Zeitschr. Kristallographie*, 2006, 221, No 5-7, 450-463.

[50] Gusev, A.I.; Rempel, A.A. Vacancy distribution in ordered Me_6C_5-type carbides. *J. Phys. C: Solid State Phys.*, 1987, 20, No 31, 5011-5025.

[51] Rempel, A.A. Atomic and vacancy ordering in nonstoichiometric carbides. *Usp. Fiz. Nauk*, 1966, 166, No 1, 33-62 (*in Russian*). (English Transl: 1996 *Physics – Uspekhi*, 1996, 39, No 1, 31-56).

[52] Rempel, A.A. Nanotechnologies. Properties and applications of nanostructured materials. *Usp. Khimii*, 2007, 76, No 5, 474-500 (*in Russian*). (English Transl: *Russian Chemical Reviews*, 2007, 76, No 5, 435-461).

[53] Gusev, A.I. Nanocrystalline materials: synthesis and properties. In: Schwarz J.A.; Contescu, C.; PutyeraK., editors. *Dekker Encyclopedia of Nanoscience and Nanotechnology*. In 5 volumes / Marcel Dekker Inc.: New York, 2004.V.3, pp 2289-2304.

[54] Gusev, A.I.; Rempel, A.A. *Nanocrystalline Materials*. Cambridge Intern. Sci. Publ.: Cambridge, 2004. 351 pp.

[55] Rempel, A.A.; Gusev, A.I. Nanostructure and atomic ordering in vanadium carbide. *Pis'ma v ZhETF*, 1999, 69, No 6, 436-442 (*in Russian*). (Engl. Transl.: *JETP Letters*, 1999, 69, No 6, 472-478).

[56] Rempel, A.A.; Forster, M.; Schaefer H-E. Positron lifetime in non-stoichiometric carbides with a *B*1 (NaCl) structure. *J. Physics: Condensed Matter*, 1993, 5, No 3, 261-266.

[57] Rempel, A.A.; Gusev, A.I. Short-range order in ordered alloys and interstitial phases. 1990 *Fiz. Tverd. Tela*, 1990, 32, No 1, 16-24 (*in Russian*). (Engl. Transl.: *Sov. Physics. - Solid State*, 1990, 32, No 1, 8-13).

[58] Rempel, A.A.; Gusev, A.I. Short-range order in superstructures 1990 *Physica status solidi* (b), 1990, 160, No 2, 389-402.

[59] Cowley, J.M. An approximate theory of order in alloys. 1950 *Phys. Rev.* 1950, 77, No 5, 669-675.

[60] Rempel, A.A.; Gusev, A.I. Relation between short-range and long-range order in solid solutions with b.c.c. and f.c.c. structures. *Physica status solidi* (b), 1985, 130, No 2, 413-420.

[61] Rempel, A.A.; Gusev, A.I. The relationship between short-range and long-range order in ordered alloys. *Fiz. Metallov i Metallovedenie*, 1985, 60, No 5, 847-854 (*in Russian*). (English Transl: *Phys. Metals Metallogr.*, 1985, 60, No 5, 11-17).

In: Niobium
Editors: M. Segers and Th. Peeters

ISBN: 978-1-62808-257-9
© 2013 Nova Science Publishers, Inc.

Chapter 3

NIOBIUM–BASED ALLOYS AS HYDROGEN PERMEABLE MEMBRANES FOR HYDROGEN SEPARATION AND PURIFICATION

H. Yukawa[1,*], ***T. Nambu***[2] **and** ***Y. Matsumoto***[3]

[1]Department of Materials Science and Engineering, Graduate School of Engineering, Nagoya University, Japan
[2]Department of Materials Science and Engineering, Suzuka National College of Technology, Japan
[3]Department of Mechanical Engineering, Oita National College of Technology, Japan

ABSTRACT

Niobium (Nb) is known to exhibit the highest hydrogen permeability among metals so it is one of the most promising metals for the advanced hydrogen permeable membranes for hydrogen separation and purification. However, Nb is very sensitive to hydrogen and when it is exposed to hydrogen atmosphere, a brittle fracture occurs due to severe hydrogen embrittlement. In order to design and develop Nb–based alloys as hydrogen permeable membrane, it is important to improve the resistance to hydrogen embrittlement. For this purpose, it is necessary to investigate the mechanical properties of Nb in hydrogen atmosphere in a quantitative way. From a series of *in–situ* SP tests, a new phenomenon of

* Corresponding author: hiroshi@numse.nagoya-u.ac.jp.

the ductile–to–brittle transition of Nb is found, which takes place as a function of hydrogen concentration. The critical hydrogen concentration, DBTC (ductile–to–brittle transition concentration) is found to be about 0.2 (H/Nb), the atomic ratio of hydrogen to Nb. On the bases of these findings, a concept for alloy design of Nb–based hydrogen permeable membrane has been proposed in order to Ssatisfy both high hydrogen permeability and strong resistance to hydrogen embrittlement. Following the concept, Nb–W–Mo ternary alloy has been successively design and developed. The designed alloy possesses about 5 times higher hydrogen permeability than currently used Pd–based alloys without showing any evidence of hydrogen embrittlement.

Keywords: Hydrogen permeable membrane, alloying effect, alloy design, hydrogen diffusivity, activation energy, hydrogen permeability, hydrogen solubility, ductile–to–brittle transition concentration (DBTC)

INTRODUCTION

Palladium (Pd) and its alloys (e.g., Pd–Ag and Pd–Cu alloys) are known as hydrogen permeable membranes and used for hydrogen separation and purification purposes [1, 2]. They are also important key materials for hydrogen production by membrane reactors [3, 4]. However, Pd is a very expensive noble metal and the hydrogen permeability through Pd–based alloy membrane is still not enough for the practical use. In addition, the mechanical properties of pure Pd and Pd–based alloy membrane are not sufficient [5]. Hydrogen permeable membranes are used under high pressure of hydrogen–containing atmosphere at high temperature. Therefore, not only the properties for functional materials (e.g., hydrogen permeability) but also the properties for high temperature structural materials (e.g., resistance to hydrogen embrittlement, high temperature strength, creep strength and phase stability at high temperature) are required.

Recently, special attention has been directed towards Nb–based alloys, because of their lower cost and higher hydrogen permeability than currently used Pd–based alloys [6-8]. For example, niobium (Nb) is inexpensive compared to Pd and exhibits the highest hydrogen permeability among metals [9]. Also Nb is one of the refractory metals, so Nb and its alloys are prospective candidate for high temperature structural materials [10, 11]. However, there is still a large barrier to the practical use due to their poor resistance to hydrogen embrittlement.

So far, there have been many attempts to improve the resistance to hydrogen embrittlement of Nb. For example, Nb–TiNi alloy (e.g., $Nb_{40}Ti_{30}Ni_{30}$ alloy) with duplex phase has been proposed which consist of the primary (Nb, Ti) and the eutectic {(Nb, Ti) + TiNi} phases [12]. It is reported that the former phase contributes mainly to hydrogen permeation, while the latter one mainly suppresses the hydrogen embrittlement. However, the microstructure of the alloy changes depending of its alloy composition, heat treatment and mechanical work (e.g., rolling), and the hydrogen permeability through the alloy membrane changes largely correspond to its microstructure. Also, the hydrogen permeability of the duplex phase alloy membrane is still not enough and almost comparable with that for Pd–based alloys [12].

Recently, the mechanical properties of pure Nb in hydrogen gas atmosphere at high temperature have been investigated by the *in–situ* small punch (SP) test method [13]. It was found that the ductile–to–brittle transition occurs drastically at the hydrogen concentration around 0.2–0.25 (H/M) at the temperature range between 573 K and 773 K without showing any micro–structural changes.

This fact suggests that the resistance to hydrogen embrittlement of Nb will be improved by keeping the hydrogen concentration less than the critical value during the operation of the practical hydrogen permeation. On the basis of these results, a concept for alloy design of Nb–based hydrogen permeable membrane has been proposed in order to satisfy both high hydrogen permeability and strong resistance to hydrogen embrittlement [14].

In the present chapter, the recent progress on the development of Nb–based hydrogen permeable membrane will be reviewed. After a brief introduction of the mechanism of hydrogen permeation through metal membrane, the mechanical properties of Nb in hydrogen atmosphere and the concept for alloy design of Nb–based hydrogen permeable alloy will be explained in detail. Then, some examples of the design and development of Nb–based alloys will be shown. Also, the alloying effects on the hydrogen diffusivity under the practical condition of hydrogen permeation through Nb–based alloy membranes are discussed.

1.1. Hydrogen Permeation through Metal Membrane

There are two kinds of hydrogen separation membranes. One is porous membranes such as polymer, zeolite, porous silica or alumina [15-17]. With these porous membranes, the selectivity (the ratio of hydrogen to other impurity gas, i.e., the purity of hydrogen) and the hydrogen flux through them (i.e., the performance) is generally incompatible. On the other hand, dense metallic membranes such as Pd–Ag and Pd–Cu alloy membranes can satisfy both of these properties [1]. There is no pure or pinhole on the membrane but only hydrogen can permeate through the membrane. Figure 1 is a schematic illustration showing the mechanism of hydrogen permeation through metal membrane which involves following several elemental steps [18]. (1) Adsorption of hydrogen molecules (H_2) on the surface. (2) Dissociation of the molecular hydrogen into atoms (H) on the surface. (3) Dissolution of atomic hydrogen into the interstitial site of the crystal lattice of metal. (4) Diffusion of hydrogen atom through the membrane. (5) Recombination of two hydrogen atoms to form hydrogen molecule on the other side of the membrane surface. (6) Desorption from the surface as molecule. These reactions occur only for hydrogen or the rate is very limited for other impurity molecules such as CO, CO_2, O_2, N_2, etc. For example, the diffusion of hydrogen atom in Nb is about 400,000 times faster than oxygen (O) atom in it at 873 K [19]. Thus, nearly perfect hydrogen permselectivity can be achieved and ultra high purity hydrogen can be obtained by a single process of hydrogen permeation with simple operation by only applying the pressure difference between the inlet and outlet sides of the membrane at high temperature.

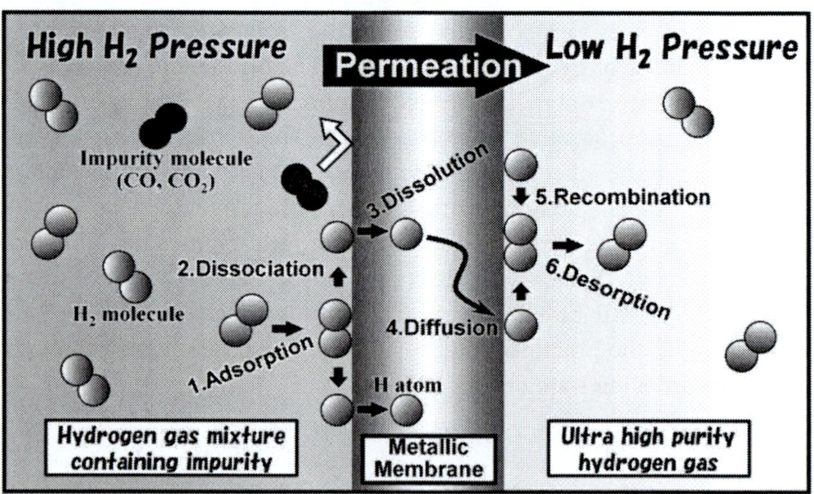

Figure 1. Schematic illustration showing the mechanism of hydrogen permeation through metal membrane.

Here, the diffusion of hydrogen atom in metal membrane (step (4)) is generally the rate–limiting process of hydrogen permeation through it. Therefore, the properties of hydrogen permeable metal membranes are commonly discussed by applying the Fick's law to the membrane with a thickness of d,

$$J = -D\frac{C_{\text{out}} - C_{\text{in}}}{d} = \frac{\Delta C}{d},$$

(1)

where J is the hydrogen flux, D is the hydrogen diffusion coefficient, C_{in} and C_{out} are the hydrogen concentration at inlet and outlet sides of the membrane. According to the Fick's law, the hydrogen flux, J, increases proportionally with increasing the gradient of hydrogen concentration, $\Delta C/d$. The hydrogen diffusion coefficient, D, is also an important factor controlling the hydrogen permeability of the metal membrane.

2. DESIGN OF NB–BASED ALLOY MEMBRANES FOR HYDROGEN SEPARATION AND PURIFICATION

2.1. Mechanical Properties of Niobium in Hydrogen at High Temperature

The hydrogen embrittlement of pure Nb has been investigated by Gahr et al., [20, 21]. The boundary for the ductile–to–brittle transition shown on the Nb–H binary phase diagram was proposed from the results of tensile test under a hydrogen gas atmosphere for the samples without any Pd coating on the surface. According to their boundary, pure niobium is supposed to be ductile in a highly soluble hydrogen state at high temperatures. However, a brittle fracture due to hydrogen embrittlement sometimes occurs during the hydrogen permeation test through a Pd–coated pure Nb metal membrane. This contradiction suggests that the hydrogen embrittlement of pure niobium has not been understood correctly as yet. Therefore, the dominant factors which cause the hydrogen embrittlement should be investigated in a quantitative way in order to design Nb–based hydrogen permeable metal membrane with strong resistance to hydrogen embrittlement.

In this study, the mechanical properties of the Nb membrane are investigated by the small punch (SP) test method which is well known as an effective method to evaluate the ductile–to–brittle transition phenomenon such as DBTT (ductile–to–brittle transition temperature) [22]. Figure 2 shows a schematic diagram of the newly developed *in–situ* SP test apparatus equipped with a gas flow system [8].

For the *in–situ* SP test, plate–shaped specimens are prepared by using a wire–electrode discharge machine. The size of the specimen is about $0.6 \times 10 \times 10 mm^3$. Both sides of the specimens are mechanically polished by emery papers followed by the final polishing with Al_2O_3 powders. The final thickness of the specimen is controlled to be 0.50 ± 0.01 mm. Subsequently, pure Pd of *ca.* 200 nm in thickness is deposited at 573 K on the surfaces by using an RF magnetron sputtering apparatus. This Pd overlayer on the surface protects the specimen from oxidation. It also acts as catalyst to eliminate the impeditive barrier for hydrogen dissociation and dissolution reactions on the specimen surface [23].

The membrane specimen is fixed at a lower flange with the die. The atmosphere between the lower and the upper flange is sealed with the bellows tube. The temperature of the specimen is controlled with the ceramic heating resistor and the thermo–couple inserted in a lower flange near the specimen.

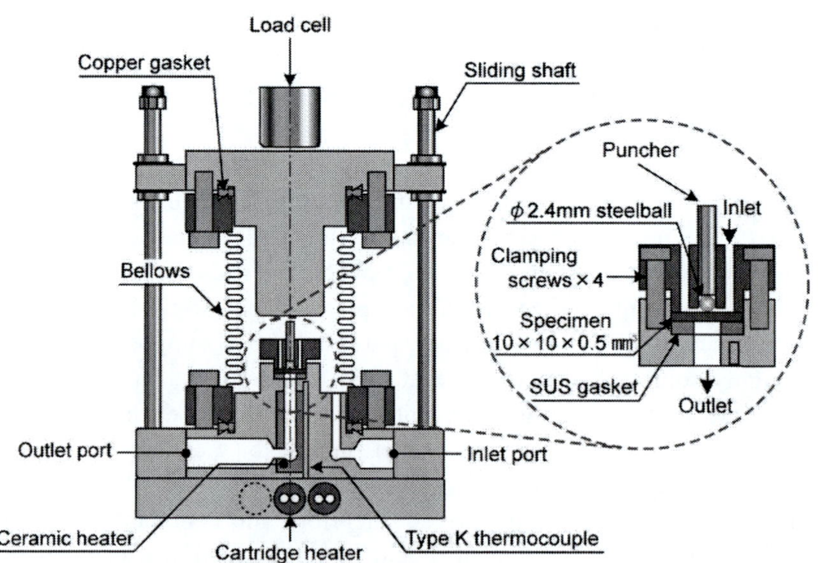

Figure 2. Schematic illustration of *in–situ* small punch (SP) test apparatus equipped with the gas flow system.

The SP rig is mounted on the Instron–type universal test machine, and then a load–deflection curve is measured by punching the specimen with a steel or Si_3N_4 ball with 2.4mm in diameter. The hydrogen pressure introduced from inlet and outlet ports can be controlled independently in this system.

The temperature and pressure conditions of the *in–situ* SP test are determined in view of the dissolved hydrogen concentration. The PCT curves of pure Nb at measured at 573K~773K by Veleckis et al. [24] or Lässer et al. [25] are used for the determination of hydrogen concentration in each specimen. For example, the conditions of the SP test conducted for pure Nb are shown in Figure 3 by numbered solid squares on each PCT curve [13]. The load–deflection curves are measured under various temperature and hydrogen pressures. These *in–situ* SP tests are performed by holding for at least 7.2ks (2hrs.) after the hydrogen gas is introduced to the SP test apparatus following each test condition. The loading rate (i.e., the cross–head speed), v, is set to be 8.3×10^{-3} mm/s (0.5mm/min). The SP absorption energies, E_{SP}, are estimated by taking the area under each load–deflection curve until the specimen failed.

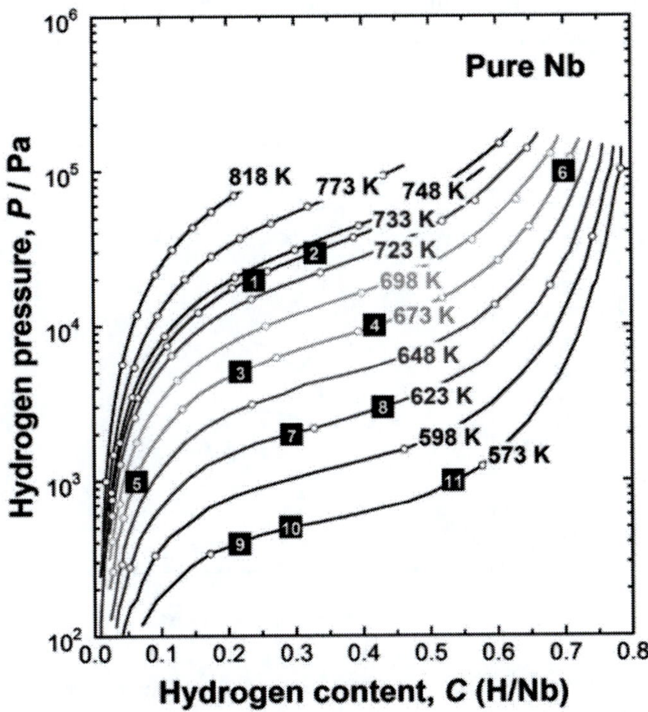

Figure 3. *In–situ* SP test conditions expressed on the PCT curves.

Some examples of the load–deflection curves of cold–rolled and annealed pure Nb metal membrane having 0.6mm of average grain size under the corresponding test conditions of Figure 3 (No. 3, 4 and 5) are shown in Figure 4 [13]. In this figure, a curve obtained in the vacuum at 673K is shown by the broken line for comparison. It is found that the maximum and failure load tend to decrease with increasing dissolved hydrogen concentration, C (H/Nb). In addition, when the *in–situ* SP tests are performed at 573K, 623K and 733K, the brittle fracture occurred under the conditions of No. 2, 4, 6, 8 and 11 shown in Figure 3, but the significant ductile fracture is observed in the test conditions of No. 3, 5, 7 and 9 shown in Figure 3, though these results does not shown in the figure.

The *in–situ* SP absorption energy, E_{SP}, is estimated from each load–deflection curve and the results are summarized in Figure 5 as a function of hydrogen concentration, C (H/Nb) [13]. As shown in the figure, the SP absorption energy for pure Nb is very small when a large amount of hydrogen dissolves in it (e.g., see inserted photo for No. 8), indicating that a severe hydrogen embrittlement occurs under higher hydrogen concentration region. However, the E_{SP} value increases rapidly with decreasing the hydrogen concentration around 0.2–0.25 (H/Nb), and typical ductile fracture morphology appears (e.g., see inserted photo for No. 9).

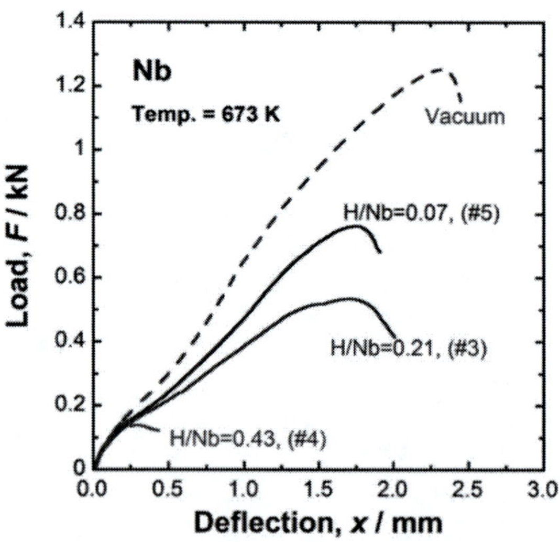

Figure 4. Load–deflection curves of Pd–coated pure Nb membrane measured by the *in–situ* SP test under the constant hydrogen pressure given in the conditions of No. 3 to 5 or in vacuum.

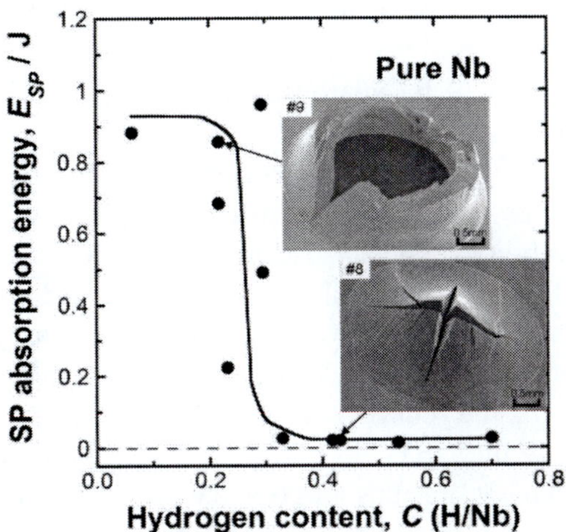

Figure 5. Change in the SP absorption energy with hydrogen concentration, C (H/Nb) for Pd–coated Nb samples. Typical fracture morphologies after SP tests are also shown in the figure; No. 8 for brittle and No. 9 for ductile manners.

The ductile–to–brittle transition occurs drastically at the hydrogen concentration around 0.2–0.25 (H/M) at the temperature range between 573K~773K. This transition does not accompanies any micro–structural changes. For example, according to the equilibrium phase diagram of Nb–H system [26] and also to the *in–situ* XRD experiments [27], no phase transition or precipitation of the secondary phase (e.g., hydride) occurs at this temperature and hydrogen pressure conditions.

Thus, the boundary of the ductile–to–brittle transition for Pd–coated Nb metal membrane is found to be shifted greatly to lower hydrogen concentration region, as compared with the ductile–to–cleavage fracture transition boundary for Pd–uncoated Nb proposed by Gahr et al. [21]. Here, it is important to note that there is still a window below the hydrogen concentration less than 0.2 (H/M), in which a ductile fracture mode is dominant.

2.2. Concept for Alloy Design of Nb–Based Hydrogen Permeable Alloy

As mentioned above, the ductile–to–brittle transition occurs drastically at the hydrogen concentration around 0.2–0.25 (H/M). This critical hydrogen concentration is referred to as the DBTC (the Ductile–to–Brittle Transition hydrogen Concentration). This fact suggests that the resistance to hydrogen embrittlement of Nb will be improved by reducing the dissolved hydrogen concentration below the DBTC.

On the basis of these results, a concept for alloy design of Nb–based hydrogen permeable membrane has been proposed which satisfies high hydrogen permeability together with strong resistance to hydrogen embrittlement [14]. A schematic illustration showing the concept is shown in Figure 6 [14]. As mentioned above, a ductile fracture takes place when the hydrogen concentration is limited less than 0.2 (H/Nb) at high temperature. The first point of the concept is to control and keep the hydrogen concentration below the DBTC in order to prevent the brittle fracture due to hydrogen embrittlement. For this purpose, the pressure–composition–isotherm (PCT) curve of pure Nb should be shifted towards the left and upper region as shown in Figure 6. In other words, the heat of hydrogen dissolution into Nb should be reduced in some ways, for example by alloying.

The second point of the concept is to optimize the pressure condition of hydrogen permeation (i.e., the hydrogen pressures at the inlet and outlet sides of the membrane) so that the concentration difference between the inlet and outlet sides of the membrane, ΔC, becomes large. For example, as shown in Figure 6, the ΔC value is larger for the designed alloy than pure Nb under the given pressure condition. In this case, higher hydrogen flux through the membrane is expected for the designed alloy than pure Nb following Eq.(1).

In addition, as mentioned before, the hydrogen permeable membrane is not only a functional material but also a structural material which need to bear up against high pressure at high temperature. Therefore, the properties for heat registrant structural material such as high temperature strength, creep strength and phase stability are very important to guarantee the long term durability at the operating temperature. For this reason, we believe that the hydrogen permeable alloys should be composed of a stable single solid solution phase with simple bcc (body–centered cubic) crystal structure.

The alloying elements are selected from two points of view. In order to reduce the hydrogen concentration, the alloying elements may need to have smaller affinity for hydrogen than Nb. The heat of hydrogen dissolution into metals has been calculated by Fukai [28]. The elements which have higher heat of hydrogen dissolution than Nb may have smaller affinity for hydrogen than Nb, and will became the candidate of the alloying elements.

On the other hand, in order to make the alloy single phase with bcc crystal structure, the alloying element should have wide solid solubility into Nb. For instance, the solubility limit of ruthenium (Ru) into Nb is about 40 mol % [26]. From these two points of view, the alloying elements are selected and the final candidates are, for example, tungsten (W), molybdenum (Mo), Ru, Pd, etc. In this paper, W and Mo are selected as alloying elements. According to the equilibrium binary phase diagrams [26], all the combinations of Nb–W, Nb–Mo and W–Mo are continuous solid solution systems.

2.3. Hydrogen Solubility

The pressure–composition–isotherm (PCT) curves are measured using a Sieverts' –type apparatus. Nb–5mol%W and Nb–5mol%W–5mol%Mo alloys are arc–melted by using a tri–arc furnace in a purified argon gas atmosphere. The purities of the raw materials used in this study are 99.96 mass% for Nb and 99.95 mass% for W and Mo. From the XRD experiments, all the samples treated in this study are confirmed to be composed of a single solid solution phase with simple bcc crystal structure. There is no evidence of the existence of any secondary phases. The PCT curves are measured at 673~773K up to about 5 MPa of hydrogen pressure.

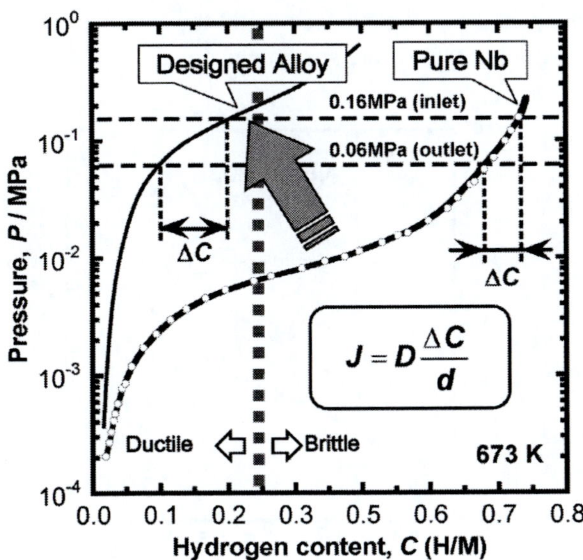

Figure 6. Schematic illustration showing the concept for alloy design of Nb–based hydrogen permeable membrane with high hydrogen permeability and strong resistance to hydrogen embrittlement.

Figure 7 shows the PCT curves for pure Nb, Nb–5mol%W and Nb–5mol%W–5mol%Mo alloys measured at 673K. The boundary of the DBTC is indicated by a broken line in the figure. It is evident that the PCT curve shifts toward left and upper region by the addition of 5 mol% of W into Nb. It further shifts by the addition of another 5 mol% of Mo into Nb–5mol%W alloy. As a result, the resistance to hydrogen embrittlement is improved by alloying of W and Mo. In fact, the applicable hydrogen pressures at the DBTC increases in the order, pure Nb < Nb–5mol%W < Nb–5mol%W–5mol%Mo.

Figure 8 shows the PCT curves of Nb–5mol%W–5mol%Mo alloy measured at 673~773K. The PCT curve further shifts toward left and upper region with increasing temperature. As a result, about 0.1 MPa of hydrogen pressure can be applied to Nb–5mol%W–5mol%Mo alloy membrane at 773K.

2.4. Hydrogen Permeability

The hydrogen permeation tests are performed at 673~773K by the conventional gas permeation method in order to evaluate the hydrogen permeability. The disk specimens of about φ12 mm in diameter with a thickness of about 0.65 mm are prepared. They are polished mechanically and pure Pd are deposited on the surface with the same procedure mentioned above. The final thicknesses of the specimens are about 0.48~0.52 mm.

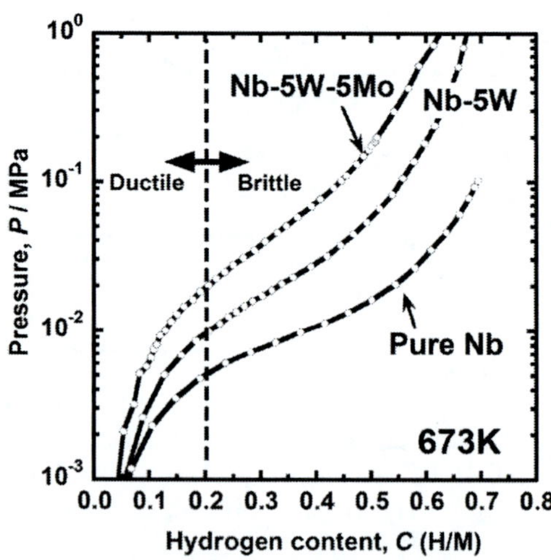

Figure 7. PCT curves for pure Nb, Nb–5mol%W and Nb–5mol%W–5mol%Mo alloys measured at 673 K.

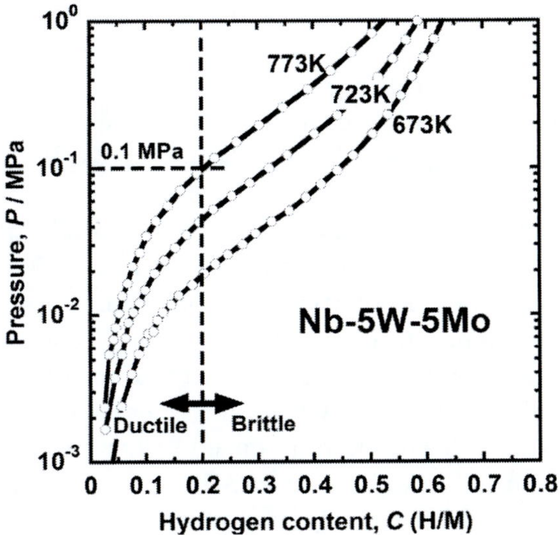

Figure 8. PCT curves for Nb–5mol%W–5mol%Mo alloy measured at 673~773 K.

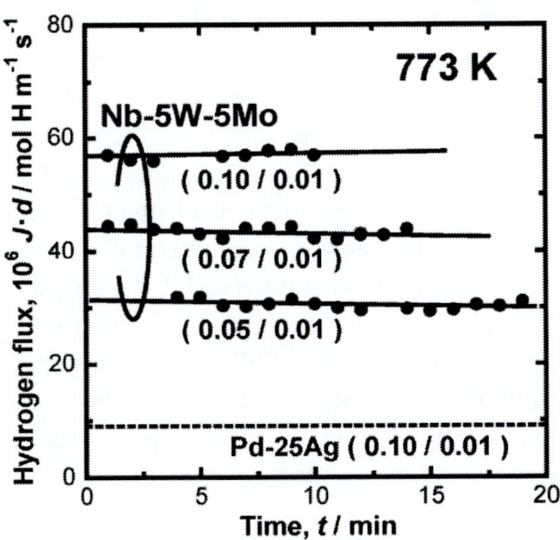

Figure 9. Time dependences of the the normalized hydrogen flux, $J \cdot d$, during measurement at 773 K for Nb–5mol%W– 5mol%Mo. The estimated value for Pd–25mol%Ag alloy is also presented in the figure. The pressure conditions are indicated in the figure as (inlet side / outlet side [MPa]).

The disk sample is set to the hydrogen permeation apparatus and then evacuated. Subsequently, it is heated up to the measuring temperature, and then a high purity (99.99999%) hydrogen gas is introduced to both sides of the disk specimen. The testing conditions of the hydrogen pressures at the inlet side are determined so that the hydrogen concentration do not exceed the critical value of the ductile–to–brittle transition, DBTC. The hydrogen pressure at the outlet side are fixed to be 0.1 MPa in this study. The hydrogen flux, J, permeated through the disk specimen is measured by a mass–flow meter.

The steady–state hydrogen flux, J, for Nb–5mol%W–5mol%Mo alloy membrane is measured. It is divided by the inverse of the sample thickness, $1/d$, in order to estimate the normalized hydrogen flux, $J \cdot d$. It is noted here that the atomic hydrogen flux, (mol H m^{-1} s^{-1}), is evaluated in this paper, which is twice as large as the gaseous hydrogen flux, (mol H$_2$ m^{-1} s^{-1}).

The results of the hydrogen permeation test for Nb–5mol%W–5mol%Mo alloy membrane at 773K are shown in Figure 9. The pressure conditions are indicated in the figure as (inlet / outlet MPa). As shown in Figure 9, the hydrogen flux is stable and nearly constant during the measurement. The estimated value of the hydrogen flux for Pd–25mas%Ag alloy under the pressure condition of (0.10 / 0.01 MPa) at 773K [29] is also indicated by a broken line in the figure for comparison. It is evident that the Nb–5mol%W–5mol%Mo alloy exhibit excellent hydrogen permeability. In fact, the normalized hydrogen flux, $J \cdot d$, for Nb–5mol%W–5mol%Mo alloy membrane measured under the pressure condition of (0.10 / 0.01 MPa) is about 5 times higher than that for Pd–25mass%Ag alloy membrane under the same pressure condition. Similar results are also obtained at other measuring temperature, 673K and 723K. From these results, the hydrogen permeation coefficient is estimated for each measuring temperature and the results are shown in Figure 10 as a function of the inverse of temperature. For comparison, the hydrogen permeability for Pd–25mass%Ag [29], V–15mol%Ni [30] and Nb–30mol%Ti–30mol%Ni [12] alloys are also presented in the figure. It is evident that the hydrogen permeability for Nb–5mol%W–5mol%Mo alloy is about 5 times higher than that for Pd–25Ag, V–15Ni and Nb–TiNi alloys at 673~773K.

2.5. Resistance to Hydrogen Embrittlement

After the hydrogen permeation test, the system is evacuated and gas leak test is performed with He. Then, the system is cooled down to room temperature in order to take out the sample from the cell to check the damage on the sample due to hydrogen embrittlement. Photo images of the disk samples after the hydrogen permeation test are shown in Figure 11(a) for Nb–5mol%W–5mol%Mo alloy and (b) for pure Nb. Brittle fracture occurs for pure Nb due to severe hydrogen embrittlement. However, for Nb–5mol%W–5mol%Mo alloy, there is an imprint made by a gasket sealing but no evidence of cracking on the sample due to hydrogen embrittlement. Thus, Nb–5mol%W–5mol%Mo alloy exhibits excellent hydrogen permeability and strong resistance to hydrogen embrittlement.

2.6. Alloying Effects on Hydrogen Diffusivity under the Practical Condition of Hydrogen Permeation

As mentioned before, the hydrogen diffusivity is an important factor controlling the hydrogen permeability. Here, it is a general idea that the hydrogen diffusion coefficients in metals change depending on the hydrogen concentration in them [31-33]. In fact, the activation energy for hydrogen diffusion in pure Nb increases with increasing the hydrogen concentration [32, 33]. Therefore, for hydrogen permeable membranes, it is important to investigate the hydrogen diffusion coefficient under the practical condition of hydrogen permeation.

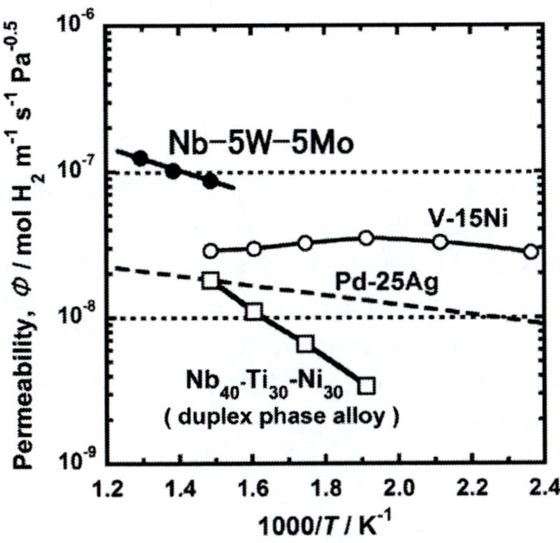

Figure 10. Comparison of the hydrogen permeability of Nb–5mol%W–5mol%Mo alloy with the values of Pd–25mass%Ag, V–15mol%Ni and Nb–30mol%Ti–30mol%Ni alloys.

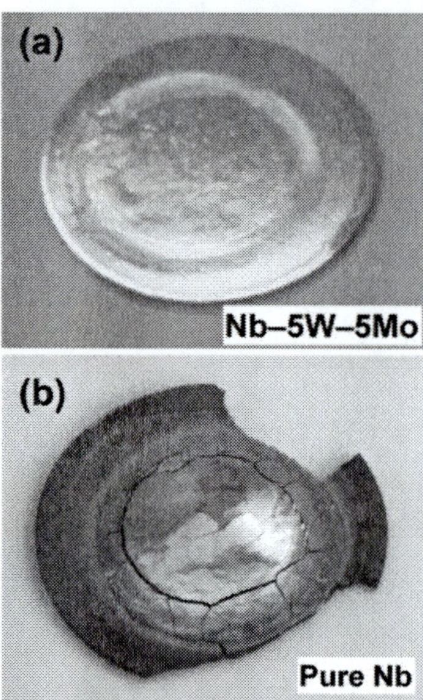

Figure 11. Appearance of the disk specimen for Nb–5mol%W–5mol%Mo alloy evacuated and cooled down to room temperature after the hydrogen permeation test.

Applying the Sieverts' law, $C = K\sqrt{P}$, to the metal–hydrogen system, the normalized hydrogen flux, $J \cdot d$, can be written as,

$$J \cdot d = D \cdot K \times (\sqrt{P_{in}} - \sqrt{P_{out}}),$$

(2)

where K is the Sieverts' constant (or the hydrogen solubility coefficient), P_{in} and P_{out}, are the inlet and the outlet hydrogen pressures, respectively. The product of D and K in Eq.(2) is defined as the hydrogen permeation coefficient, $\phi = (D \cdot K)$ and generally used as a measure to evaluate the performance of the hydrogen permeable alloys.

Then, the hydrogen diffusion coefficient during the hydrogen permeation will be estimated from the following relationship,

$$D = \frac{\phi}{K} = \frac{J \cdot d}{K \times (\sqrt{P_{in}} - \sqrt{P_{out}})},$$

(3)

by only measuring the hydrogen flux, J, if the Sieverts' constant, K, is known. Such analysis of the hydrogen diffusion coefficients have commonly been performed for hydrogen permeable metal membranes, especially for Pd–based alloys, because the hydrogen diffusion coefficient under the condition of hydrogen permeation can be estimated easily without measuring the hydrogen concentration gradient, $\partial C/\partial x$, across the membrane. However, this easy and convenient method cannot be applied to the system where the Sieverts' law is not satisfied at the hydrogen pressures for hydrogen permeation. This is because the Sieverts' constant, K, cannot be determined uniquely for such metal–hydrogen system under the practical condition of hydrogen permeation. For example, the PCT curves for Pd–26mol%Ag alloy and pure Nb measured at 773K [34] are shown in Figure 12. Here, the square root of hydrogen pressure is plotted as a function of hydrogen concentration. As shown in Figure 12(a), the Sievelts' law is almost satisfied for Pd–Ag system. On the contrary, for Nb–H system shown in Figure 12(b), the Sievelts' law is only fulfilled at low hydrogen pressures below 0.003 MPa, and no longer valid at the pressure range for the practical hydrogen permeation.

In this study, the hydrogen diffusion coefficients under the practical conditions of hydrogen permeation have been estimated following Eq.(1). The normalized hydrogen fluxes, $J \cdot d$, are analyzed in view of the hydrogen concentration difference, ΔC, between the inlet and outlet sides of the disk sample. Here, the ΔC values are estimated from the PCT curves shown in Figure 8.

The correlation between the normalized hydrogen flux, $J \cdot d$, and the difference of hydrogen concentration, ΔC, are shown in Figure 13 for Nb–5mol%W–5mol%Mo alloy. Here, the lattice expansion due to thermal heating [35, 36] and hydrogen uptake into metals [37] are taken into account in order to estimate the ΔC value in the unit of hydrogen volumetric molar concentration, (mol H m^{-3}), instead of the atomic ratio, (H/M).

As shown in Figure 13, there is a linear relationship between the normalized hydrogen flux and the concentration difference for each measuring temperature. Each straight line shown in Figure 13 crosses at the origin, indicating that the diffusion–limiting hydrogen permeation reaction takes place through the membrane following the Fick's law. Then, the hydrogen diffusion coefficients during hydrogen permeation are evaluated from the slope of the lines shown in Figure 13.

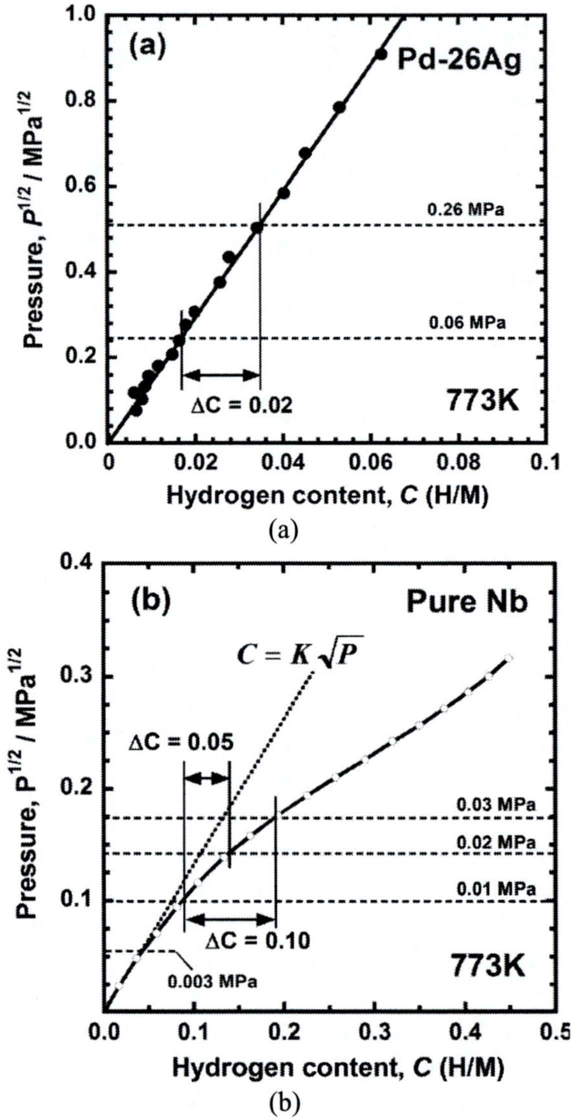

Figure 12. PCT curves for (a) Pd–26mol%Ag alloy and (b) pure Nb measured at 773 K.

Figure 14 shows the Arrhenius plots of the hydrogen diffusion coefficient during hydrogen permeation. For comparison, the hydrogen diffusion coefficient for pure Nb and Nb–5mol%W alloy estimated by the same permeation method [38] are presented in the figure. It is found that the hydrogen diffusion coefficient under the practical condition of hydrogen permeation increases by the addition of W and Mo into Nb, especially at low temperature.

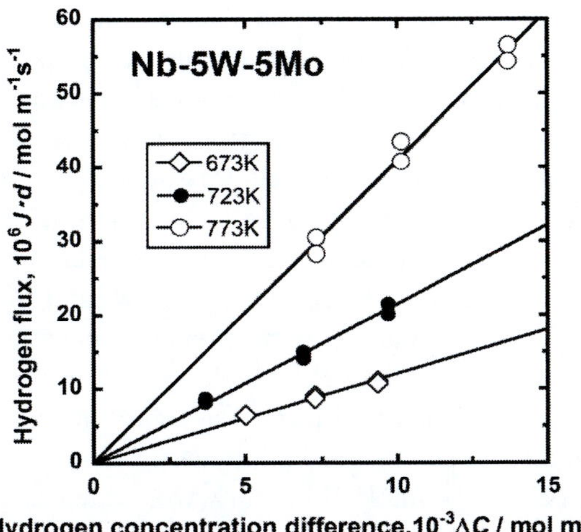

Figure 13. Correlation between the normalized hydrogen flux and the hydrogen concentration difference for Nb–5mol%W–5mol%Mo alloy at 673~773K.

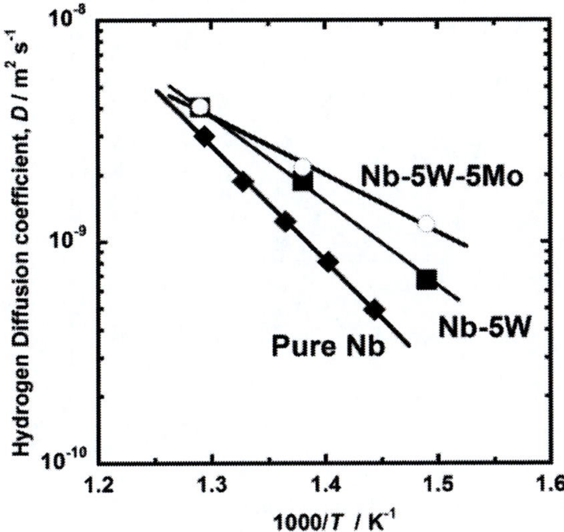

Figure 14. Arrhenius plots of the hydrogen diffusion coefficient under the condition of hydrogen permeation for pure Nb, Nb–5mol%W and Nb–5mol%W–5mol%Mo alloys.

The activation energy for hydrogen diffusion are estimated from the straight line shown in Figure 14. They are 98.6 kJ/mol, 77.9 kJ/mol and 50.7 kJ/mol for pure Nb, Nb–5mol%W and Nb–5mol%W–5mol%Mo alloys, respectively. It is found that the addition of W and Mo into Nb decreases the activation energy for hydrogen diffusion. This may be understood in terms of the distribution of the chemical potential of hydrogen atom in the alloys. From the results of PCT measurements shown in Figure 7, the addition of W and Mo into Nb increases the equilibrium hydrogen pressures at the same hydrogen concentration. For example, the hydrogen pressures at the DBTC at 673 K are about 0.005 MPa, 0.01 MPa and 0.02 MPa for pure Nb, Nb–5mol%W and Nb–5mol%W–5mol%Mo alloys, respectively. Therefore, the stability of hydrogen atom in Nb decreases by the addition of W and Mo. In other words, as shown in Figure 15 schematically, the chemical potential of the hydrogen atom at the interstitial site of the bcc crystal lattice in Nb become shallow by alloying of W and Mo, which will lead to decrease the activation energy for hydrogen diffusion. In fact, the activation energy changes in the order, pure Nb > Nb–5mol%W > Nb–5mol%W–5mol%Mo, which is the same order of the change in the equilibrium hydrogen pressure. Thus, the designed Nb–5mol%W–5mol%Mo alloy exhibits enhanced hydrogen diffusivity for hydrogen permeation.

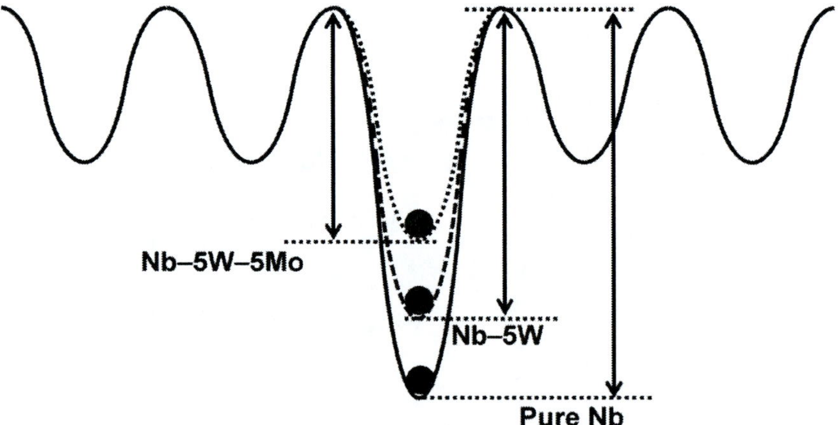

Figure 15. Schematic illustration showing the chemical potential of hydrogen atom at the interstitial site in pure Nb and Nb–based alloys.

CONCLUSION

From a series of the *in–situ* SP tests, the critical hydrogen concentration to the ductile–to–brittle transition (DBTC) for pure Nb is found to be about 0.2 (H/M). These results suggest that Nb is still ductile even in hydrogen atmosphere if the dissolved hydrogen concentration is less than the critical value of the DBTC. On the basis of these results, a concept for alloy design of Nb–based hydrogen permeable membrane has been proposed. The concept is applied to the design of Nb–W–Mo ternary system, and Nb–5mol%W–5mol%Mo alloy membrane has been successfully design and developed which exhibits excellent hydrogen permeability together with strong resistance to hydrogen embrittlement. Also, the addition of W and Mo into Nb is found to enhance the hydrogen diffusivity under the practical condition of hydrogen permeation.

REFERENCES

[1] Paglieri, S. N. & Way, J. D. (2002). "Innovations in palladium membrane research", *Sep. Purif. Methods*, *31*, 1–169.

[2] Mordkovich, V., Baichtock, Y. K. & Sosna, M. (1992). "The large scale production of hydrogen from gas mixtures: A use for ultra thin palladium alloy membranes", *Int'l J. Hydrogen Energy*, *18*, 539-544.

[3] Chen, Y., Wang, Y., Xu, H. & Xiong, G. (2008). "Efficient production of hydrogen from natural gas steam reforming in palladium membrane reactor", *Applied Catalysis B: Environmental*, *81*, 283-294.

[4] Tokyo Gas Co., Ltd. et al. (2008). *"Development of the hydrogen production system using a membrane reformer"*, NEDO (FY2005–FY2007) Final Report, Project No. P03015.

[5] Watanabe, N., Yukawa, H., Nambu, T., Matsumoto, Y., Zhang, G. X., & Morinaga, M. (2009). "Alloying effects of Ru and W on the resistance to hydrogen embrittlement and hydrogen permeability of niobium", *J. Alloys Compd.*, *477*, 851-854.

[6] Buxbaum, R. E. & Kinney, A. B. (1996). "Hydrogen Transport through Tubular Membranes of Palladium–Coated Tantalum and Niobium", *Industrial & Engineering Chemical Research*, *35*, 2, 530-537.

[7] Hashi, K., Ishikawa, K., Matsuda, T. & Aoki, K. (2004). "Hydrogen permeation characteristics of multi–phase Ni–Ti–Nb alloys", *J. Alloys. Compd.*, *368*, 215–220.

[8] Nambu, T., Shimizu, K., Matsumoto, Y., Rong, R., Yukawa, H., Morinaga, M. & Yasuda, I. (2007). "Enhanced Hydrogen Embrittlement of Pd–coated Niobium Metal Membrane Detected by In–situ Small Punch Test under Hydrogen Permeation", *J. Alloys Compd.*, 446–447, 588–592.

[9] Steward, S. A. (1983). *"Review of Hydrogen Isotope Permeability Through Materials"*, Lawrence Livemore National Laboratory Reports, UCRL–53441.

[10] Tan, Y., Tanaka, H., Ma, C., Kasama, A., Tanaka, R., Mishima, Y. & Hanada, S. (2000). "Solid–Solution Strengthening and High–Temperature Compressive Strength of Nb–X Alloys (X=Ta, V, Mo and W)", *J. Jpn. Inst. Metals*, *64*, 559-565.

[11] Inoue, S., Kato, M., Kano, S.,Isshiki, Y., Saito, J.,Yoshida, E. & Morinaga, M. (1994). "Design of Super Heat–Resisting Nb–Based Alloys for Nuclear Applications", *J. Jpn. Inst. Metals*, *58*, 826-834.

[12] Ishikawa, K., Tokui, S. & Aoki, K. (2009). "Microstructure and hydrogen permeation of cold rolled and annealed $Nb_{40}Ti_{30}Ni_{30}$ alloy", *Intermetallics*, *17*, 109–114.

[13] Matsumoto, Y., Yukawa, H. & Nambu, T. (2010). "Quantitative Evaluation of Hydrogen Embrittlement of Metal Membrane Detected by in–situ Small Punch Test under Hydrogen Permeation", *Metall. J.*, *LXIII*, 74–78.

[14] Yukawa, H., Nambu, T., Matsumoto, Y., Watanabe, N., Zhang, G. X. & Morinaga, M. (2008). "Alloy Design of Nb–Based Hydrogen Permeable Membrane with Strong Resistance to Hydrogen Embrittlement", *Mater. Trans.*, *49*, 2202–2207.

[15] Gavalas, G. R., Megiris, C. E. & Nam, S. W. (1989). "Deposition of H_2–permselective SiO_2 films", *Chem. Eng. Sci.*, *44*, 1829-1835.

[16] Niwa, M., Kato, S., Hattori, T. & Murakami, Y. (1984). "Fine control of the pore–opening size of the zeolite mordenite by chemical vapour deposition of silicon alkoxide", J. *Chem. Soc., Farad. Trans.*, *80*, 3135-3145.

[17] Cheng, Y. S., Peña, M. A., Fierro, J. L., Hui, D. C. W. & Yeung, K. L. (2002). "Performance of alumina, zeolite, palladium, Pd–Ag alloy membranes for hydrogen separation from Towngas mixture", *J. Membr. Sci.*, *204*, 329-340.

[18] Yukawa, H., Nambu, T. & Matsumoto, Y. (2013). "Design and development of Nb–W–Mo alloy membrane for hydrogen separation and purification", *Defect and Diffusion Forum*, *333*, 61-71.

[19] Peterson, N. L. (1960). *Diffusion in Refractory Metals*, WADD Technical Report 60-793.

[20] Gahr, S., Grossbeck, M. L. & Birnbaum, H. K. (1977). "Hydrogen embrittlement of Nb I—Macroscopic behavior at low temperatures", *Acta Metall.*, *25*, 125-134.

[21] Gahr, S. & Birnbaum, H. K. (1978). "Hydrogen embrittlement of niobium—III. High temperature behavior", *Acta Metall.*, *26*, 1781-1788.

[22] Baik, J.M., Kameda, J. & Buck, O. (1986). "Development of Small Punch Tests for Ductile–Brittle Transition Temperature Measurement of Temper Embrittled Ni–Cr Steels", in Corwin, W. R. and Lucas, G. E. eds., *ASTM STP 888*, Philadelphia, 92-110.

[23] Amano, M., Komaki, M. & Nishimura, C. (1991). "Hydrogen permeation characteristics of palladium–plated V–Ni alloy membranes", *J. Less-Common Met.*, 172-174, Part 2, 727-731.

[24] Veleckis, E. & Edwards, R. K. (1969). "Thermodynamic properties in the systems vanadium–hydrogen, niobium–hydrogen, and tantalum–hydrogen", *J. Phy. Chem.*, *73*, 683-692.

[25] Lässer, R., Meuffels, P. & Feenstra, R. (1988). "Datenbank der Löslichkeiten der Wasserstoffisotope Protium (H). Deuterium (D) und Tritium (T) in den Mtallen V, Nb, Ta Pd und den Legierungen $V_{1-x}Nb_x$, $V_{1-x}Ta_x$, $Nb_{1-x}Mo_x$, $Pd_{1-x}Ag_x$", Berichte der Kernforschungsanlage Jülich 214 S., JUEL-2183.

[26] Handbook of Binary Alloy Phase Diagrams, ASM International CD version 1.0, ISBN 0-87170-562-1.

[27] Nambu, T., Shimizu, N., Ezaki, H., Yukawa, H., Morinaga, M. & Takeichi, N. (2006). "In–Situ X–ray Diffraction Measurement for Pure Niobium Metal in High Temperature Hydrogen Atmosphere", *J. Jpn. Inst. Metals*, *70*, 467-472 (in Japanese).

[28] Fukai. Y. (1991). "Properties of Hydrogen in Metal Lattice", *Shokubai (Catalysts & Catalysis)*, *33*, 254–260 (in Japanese).

[29] Serra, E., Kemali, M., Perujo, A. & Ross, D. K. (1998). "Hydrogen and Deuterium in Pd–25 pct Ag Alloy: Permeation, Diffusion, Solubilization, and Surface Reaction", *Metall. Mater. Trans. A, 29A*, 1023–1028.

[30] Nishimura, C. Komaki, M., Hwang, S. & Amano, M. (2002). "V–Ni alloy membranes for hydrogen purification", *J. Alloys Compd., 330–332*, 902–906.

[31] Bauer, H. C., Völkl, J., Tretkowski, J. & Alefeld, G. (1978). "Diffusion of hydrogen and deuterium in Nb and Ta at high concentrations", *Z Physik B 29*, 17-26.

[32] Zojal, O. J. & Cotts, R. M. (1975). "Self–diffusion coefficient of hydrogen in $NbH_{0.6}$", *Phys. Rev. B, 11*, 2443-2446.

[33] Mauger, P. E., Williams, W. D. & Cotts, R. M. (1981). "Diffusion and NMR spin lattice relaxation of ^1H in a' TaH_x and NbH_x", *J. Phys. Chem. Solids, 42*, 821-826.

[34] Yukawa, H., Zhang, G. X., Watanabe, N., Morinaga, M., Nambu, T. & Matsumoto, Y. (2009). "Analysis of hydrogen diffusion coefficient during hydrogen permeation through niobium and its alloys". *J. Alloys Compd., 476*, 102-106.

[35] Davis, J. R. (Ed.). (1990). *Metals Handbook*, 10th edn., Vol. *2*, ASM International, Metals Park, OH., ISBN 0-87170-378-5.

[36] Asaoka, K. & Kuwayama, N. (1990). "Temperature Dependence of Thermal Expansion Coefficient for Palladium–based Binary Alloy", *Dental Mater. J., 9*, 47-57.

[37] Alefeld, G. & Völkl, J. (1978). "Hydrogen in metals I − Basic properties", *Topics in Applied Physics, 28*. ISBN 3-540-08705-2.

[38] Zhang, G. X., Yukawa, H., Nambu, T., Matsumoto, Y., & Morinaga, M. (2010). "Alloying effects of Ru and W on hydrogen diffusivity during hydrogen permeation through Nb–based hydrogen permeable membranes", *Int'l J. Hydrogen Energy, 35*, 1245-1249.

In: Niobium
Editors: M. Segers and Th. Peeters

ISBN: 978-1-62808-257-9
© 2013 Nova Science Publishers, Inc.

Chapter 4

ELECTRODE PROCESSES IN ANODIC OXIDE FILMS OF NIOBIUM

Leonid Skatkov[1,*] *and Valeriy Gomozov*[2,†]
[1]PCB "Argo", Beer Sheva, Israel
[2]NTU "Kharkov Polytechnical Institute", Kharkov, Ukraine

ABSTRACT

We review the following issues: peculiarities of polycrystalline AOP development at niobium in nitrate salt melt at temperatures of Nb_2O_5 recrystallization, the mechanisms of change in structure and properties of niobium pentoxide in the result of electrode processes under cathodic polarization (at the model structure of polycrystalline film), as well as the methods of amorphous electrochromic AOP development and ways of electrochromic processes optimization at such films.

INTRODUCTION

Anodic oxide films (AOF) of niobium Nb_2O_5 are widely used for generation of solid-electrolyte capacitors (SEC), which implement electric capacitance in «Nb – AOF Nb_2O_5 – electrolyte» system (MOE – system), and

---

[*] 4/23 Shaul ha-Melekh Street, 84797 Beer Sheva, Israel.
[†] 21 Frunze Street, 61002 Kharkov, Ukraine.

electrochromic indicating devices (EID), operating due to electrochromic effect (ECE) – reversible change of optical and electrophysical properties of Nb_2O_5 films at alternate electrochemical polarization in MOE-systems.

It shall be admitted that if in the case of SEC exclusively amorphous AOF (because of their high dielectric properties) are used, and the development of such films have been worked out in detail. Alternatively, ECE has originally been discovered at polycrystalline AOF, hardly suitable for device applications because of the relaxation phenomena of AOF kinetics, caused due to heterogeneity of polycrystalline structure. At the same time, as it will be shown below, polycrystalline electrochromic AOF Nb_2O_5 are of considerable interest as a model system for studying of structural changes in AOF at conditions of cathodic polarization (CP). Understanding of mechanism of the processes occurring at the surface and in the bulk of AOF at CP is what opens new strategies for development of high-ECI and expansion of niobium SEC application range.

We review the following issues: peculiarities of polycrystalline AOP development at niobium in nitrate salt melt at temperatures of Nb_2O_5 recrystallization (Section A), the mechanisms of change in structure and properties of niobium pentoxide in the result of electrode processes under cathodic polarization (at the model structure of polycrystalline film, Section B), as well as the methods of amorphous electrochromic AOP development and ways of electrochromic processes optimization at such films (section C).

PART A

Formation features and structure of anodic oxide films (AOF) on Nb are studied in the range of papers however most of them are devoted to amorphic oxide layers [1, 2]. At the same time polycrystal niobium AOFs of 1 mkm width, which are subject to be applied as catalysts for a range of chemical reactions, adsorbers of residual gases in vacuum etc., are also worth attention. At the same time, MOE-system, "Nb – polycrystalline AOF Nb_2O_5 –salt melt KNO3" is considered as extremely useful model for analysis of electrode processes at AOF at conditions of cathodic polarization. It has really been determined that the single ion-exchange process at the oxide / electrolyte interface in this case is oxygen injection from oxide anions.

The present paper is considered as an extension of earlier investigations of niobium anodized in nitrate melts of salts at oxide recrystallization temperatures. Studied patterns of growth and crystallization of oxide layers are currently subject to be completed with investigations of AOF phase and elemental composition considering the available general data on applicability of the process of metal substrate oxygen saturation and inclusion of electrolyte components to oxide at anodic oxidation [3].

The investigation is concerned with AOF, created by anodic oxidation of niobium substrate of 0.5 mm width in KNO_3 melt at the temperature of 670K in galvanostatic mode (at current density of 1 A/cm^2) up to voltage of 25V.

Examination of phase composition applying Auger spectroscopy at layer-by-layer etching with Argon ions has shown (Figure 1) that generated films instead of being entire niobium pentoxide layer, as it was seemed to be, has occurred to be a «sandwich» of Nb_2O_5, NbO_2 and NbO phases (in the direction from oxide surface to niobium). This is similar to what is taking place at thermal oxidation [4]; at the same time within anodic oxidation oxygen is intensively dissolved in niobium substrate, penetrating into it on the depth which substantially exceeds the oxide thickness.

The last fact is conditioned due to high getteric properties of niobium towards oxygen at the temperature of 600 − 800K [4], and may considerably affect mechanical and electrophysical features of generated structures because of adjustment of metal/oxide section border.

Surveys, carried out in K α-radiation of chromic anode have contributed to appointment of the top layer of oxide film as low-temperature; α - Nb_2O_5 modification with pseudo-hexagonal screen and the following properties of elementary cell: a=0.3607 nm, c=0.3925 nm which coincides with the results of earlier surveys.

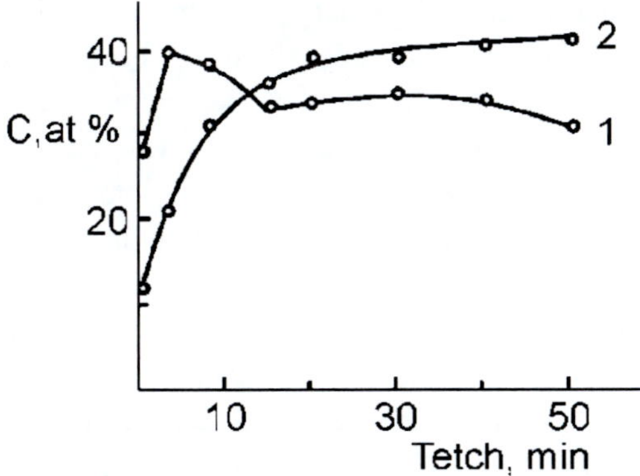

Figure 1. Clusteric sections of «niobium-AOF niobium» structures (etching velocity with Argon ions is considered to be constant at the whole width): C- concentration (at%); T_{etch} − time of etching (min); 1 − oxygen; 2 − niobium.

Studying of the possibility of electrolyte components being included into oxide has lead to a result which from the first sight may be considered as contradictory: at growth anodic oxide is saturated not only with anions (which could have been expected similar to anodic oxidation in hydrous electrolytes), but also with cations.

The Figure 3 shows clusteric distribution sections of additive anionic and cationic components in niobium AOFs. It can be seen that nitrogen concentration in the film is the highest in the adjasent to niobium layer and decreases in the direction to AOF outer surface, while potassium clusteric section has the opposite nature. Such results and the fact that aggregated concentration of potassium and nitrogen remains constant at the whole oxide width can be explained as follows.

As it has been mentioned in [5], the vital condition for initial stages of AOF growth is maintenance of constant field voltage in oxide up to the value of 20V; at the same time oxide film, remaining amorphic, includes anions, probably as to the mechanism applicable for hydrous electrolytes. This explains preferential allocation of anions in adjasent to niobium layers.

At oxide crystallization intensive AOF growth is observed. It proceeds almost without applied voltage increase. Surface dispersion occurred due to AOF crystallization substantially increases its reactivity as to the known mechanism, which together with considerable reduction of field voltage contributes to possibility of proceeding of the competing process – chemical additional oxidation of AOF by electrolyte components among which cation peroxides are considered as the most active. Consequently formation of cation peroxide solid compounds with niobium oxides becomes possible and anodic film layers adjasent to electrolyte occur to be enriched with potassium.

Considering the fact that the issue of cations presence in AOF is of great interest, potassium quantitative content in oxide has additionally been inspected by the method of fluorescent X-ray test as to intensity of K α-fluorescent X-rays excluding background. As for potassium model KCl crystal has been applied. Assuming 50% potassium content in the model and linear dependence of radiation intensivity from concentration of the element it has been calculated potassium concentration value constituting the rate close to 9%, which together with the abovementioned data of X-ray diffractometry survey allows to point on formation of potassium solid compounds in niobium oxide at AOF crystallization.

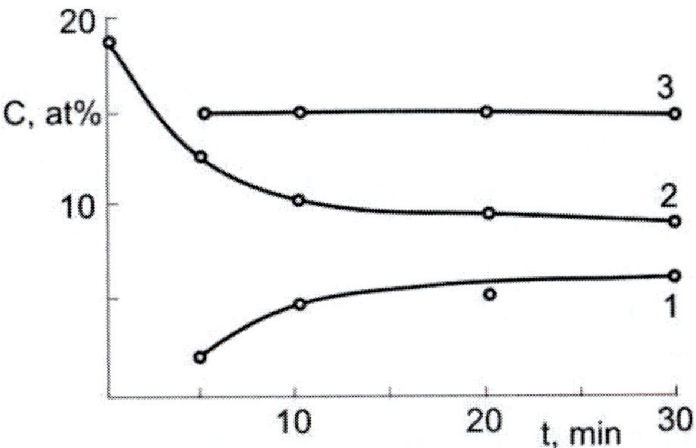

Figure 2. Clusteric sections of AOF on niobium structures (etching velocity with Argon ions is considered to be constant at the whole width): C- concentration (at%); T_{etch} – time of etching; 1 – nitrogen, 2 – potassium, 3 – overal concentration of nitrogen and potassium.

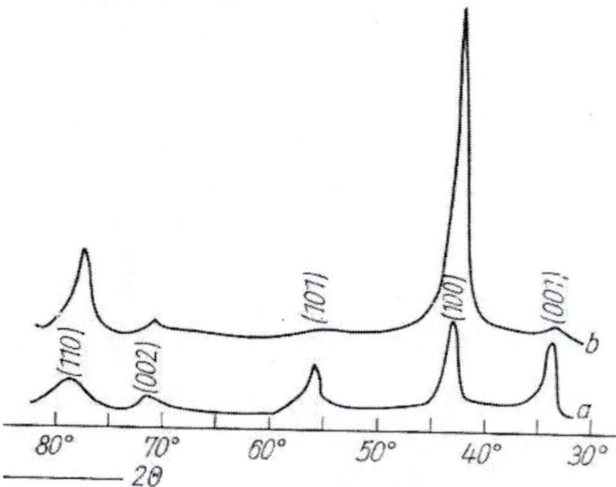

Figure 3. Structural changes in Nb_2O_5: a – initial state; b – after cathodic polarization (Θ - scattering angle).

PART B

Let us consider mechanisms of changes in structure and properties of pentoxide niobium AOF caused due to electrode processes on Nb_2O_5 at conditions of cathodic polarization of the latter in MOE system, described in the Section A.

Changes in the diffraction pattern of the Nb_2O_5 film are shown in Figure 4. As seen in the Figure, the type of the Nb_2O_5 crystal lattice remains unchanged, but one can observe f pronounced change in the relative intensity of the diffraction maxima due to disordering in the oxide [6].

The diffraction pattern (Figure 1, b) contains no lines corresponding to the lower oxide of Nb, which confirms the statement oxygen anions are extracted only from a thin near-surface layer of the AOF which, according to the Auger electron spectroscopy data, does not exceed 0.4 to 0.5 nm; the complete reversi bility of the diffraction paterns also allows to make a conclusion on the disordering of the Nb_2O_5 structure. As seen in Table 1, the anion extraction causes an anisotropic change of the Nb_2O_5 parameters, viz. enlarging of the a-axis and shortering of the c-axis. Note that, according to [7], the thermal reduction of Nb_2O_5 leads to a regular decreasing of all parameters of the oxide unit cell. The authors of [7] used single crystal samples of Nb_2O_5, which unfortunately does not allow a correct interpretation of the resulting mismatch because of the above-given consideration.

Table 1. Structural changes in Nb_2O_5

	a (nm)	c (nm)	V ($10^{-3}nm^3$)
initial state	0.36158	0.39355	44.53
intermediate state	0,36254	0.39198	44.64
maximum reduction	0.36353	38894	44.57

Along with the structural distortions, the AOF properties, viz. the spectral dependence of the film reflection coefficient $R = R(\lambda)$ (see Figure 5) and the electric conductivity, also undergo a change.

At a first glance, the most obvious reason for the change in AOF electrophysical and optical properties is the extraction of oxygen anions from the sample which compensates the injection of electrons from the metal to the oxide; these electrons stick to metal cations, thus diminishing the valency of the latter while the electric conductivity of the oxide grows with the concentration of low-valency cations [8] However, as was shown by X-ray studies of samples reduced by heating in vacuum and polarized in the melt, such samples, having practically the same form of the $R=R(\lambda)$ dependence (Figure 5 c, d), lose more oxygen, which in this case leads to formation of a new crystallographic modification. It is also difficult to explain the observed dynamics of the change in the Nb_2O_5 unit cell volume only by the oxygen extraction. Indeed, in the initial state of anion extraction the unit cell volume somewhat increases and only later, while approaching the maximum degree of the reduction, it starts decreasing. As seen in Figure 5b, the initial extraction state being accompanied by a change in the shape of the oxide reflection spectrum, virtually causes no change in the reflection coefficient in the visible range, but further on (Figure 5 c) an abrupt change in R after a slight shift of the absorption edge is observed/ At the net stage one can observe a transition of the film to a state when its colour changes visually and the Nb_2O_5 band gap width narrows to values not higher than 1 eV, while the electrical conductivity grows by the factor of 10^4.

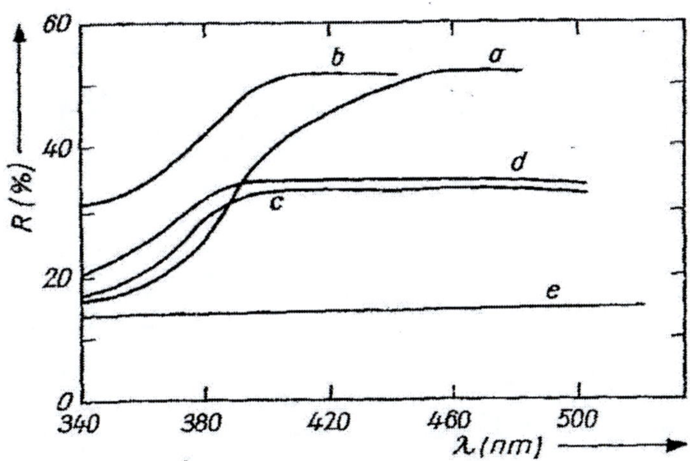

Figure 4. Spectral dependences of the oxide film reflection coefficient: a – initial state; b – beginning of anion extraction; c – intermediate state; d – thermal reduction; e – maximum reduction.

Table 2. Donor concentrations and flat-band potential in Nb₂O₅

	donor concentration (cm^{-3})		U_{FB}
	N_1	N_2^+	
initial state	2.9×10^{17}	9.1×10^{17}	-0.38
Intermediate state	3.8×10^{17}	2.2×10^{17}	-0.25
Maximum reduction	2.9×10^{20}	3.1×10^{20}	-0.57

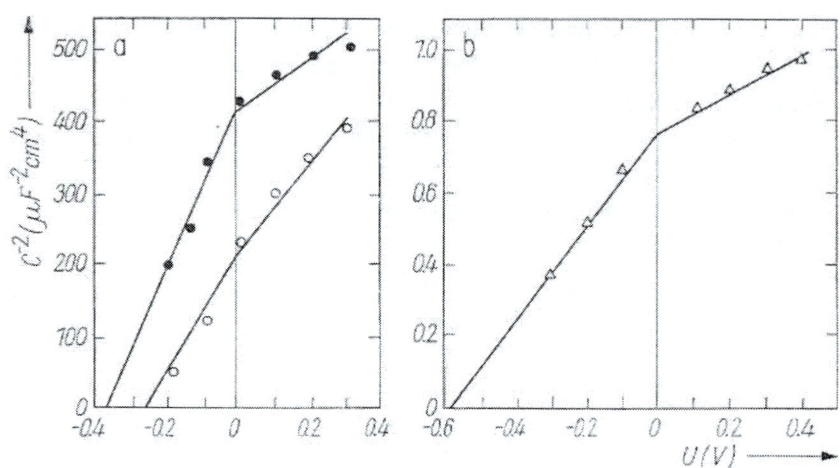

Figure 5. C^{-2} - φ graphs for Nb_2O_5: a ● initial state; ○ beginning of anion extraction; b maximum reduction.

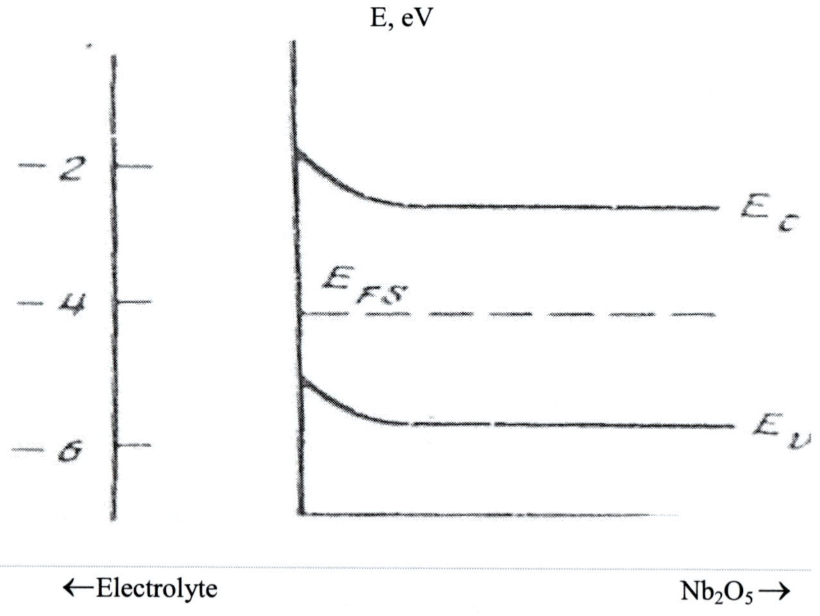

Figure 6. Position of the energy bands at the Nb_2O_5 surface.

The above-described changes in electrophysical and optical properties of Nb_2O_5 films can be interpreted as follows. The initial stage of the oxygen anion extraction results in structural changes leading to an enlarged Nb_2O_5 unit cell (due to the anisotropic change in the lattice parameters a and c) and to the appearance of the first traits of disordering, which leads to a change in the spectrum shape and, therefore, to a change in the AOF reflection coefficient. Further multiplication jf defects results in a lower disordering energy [9] and, consequently, to the transition of the AOF structure to a disordered state. The resulting distortion of the crystal lattice brings about a narrowing of the Nb_2O_5 band gap which, in its turn, results in a phase transition with a change in conductivity revealed by an abrupt change in the AOF electrophysical and optical properties, accompanied by structural changes leading to the diminishing the AOF unit cell volume. Thus, the extraction of oxygen anions from the AOF near-surface layer brings about a phase transition with a change in conductivity; the cause is a structural distortion due to disordering in Nb_2O_5..

Let us consider now the mechanism of the interaction of the oxide surface with its volume which "ensures" the change in the Nb_2O_5 properties as a result of the near-surface anion extraction.

Figure 6 shows the C – U characteristics of Nb_2O_5 at the initial state and at various stages of anion extraction given in Mott-Schottky coordinates. In accordance with [10] two linear segments with different slopes are determined by the two available types of donor centres in the oxide situated at at different depths: fully ionized (at room temperature) "shallow" donors and the partly ionized "deep" donors with the concentration N_1 and N_2 respectively. The concentration of the "deep" donors ionized in the semiconductor bulk is N_2^+ ($N_2^+ = rN_2$, r being the ionization level). Table 2 presents the concentration N_1 and N_2^+ calculated from the slopes of the liner segments by the Mott-Schottky equation [11].

The growth of the shallow donor concentration and the narrowing of the band gap, E_{gs}, on the oxide surface registered by the Nb_2O_5 flat-band potential shift, U_{FB}, towards positive values confirm that at the initial stage of the extraction the phase transition in the near-surface AOF layer follows the above-described mechanism. At the same time, the electrons injected from the metal to the oxide during the cathode polarization are trapped by the deep donor centres inside the film, thus diminishing the concentration of the ionized deep levels. The growth of the carrier concentration in the space charge region (SCR) during the phase transition leads to ionizing the deep centres in the electric fields SCR [12]; as a result, the ionized donor concentration N_2^+ becomes sufficient to carry out the phase transition in the AOF bulk (the U_{FB} shift towards negative values in the course of the further anion extraction is caused by the Nb_2O_5 crystal lattice distortion after formation of the high-depletion layer on the oxide surface, see Figure 7).

Note that above-described peculiarities of the phase transition in Nb_2O_5 are qualitatively similar to the effects observed in the metal-semiconductor phase transition in Ti_2O_3 that occurs in heating the oxide. The latter transition is smooth and does not violate the crystal lattice symmetry, although in the transition one observes an anomalous change in the c/a ratio (the c-axis increases while a-axis decreases). As was shown in [13] this phase transition can be explained by the band overlap model, in particular, by a growth of the valence band energy which is caused by changes in the metal atom separations along the c-axis.

It may be conjectured that a qualitatively similar mechanism also works in the Nb_2O_5 film. The authors of [14], using cluster model methods, suggested a band structure for Ta_2O_5 whose low-temperature modification is isostructural with Nb_2O_5. As was shown in [14], the valence band width of the oxide film depends on a parameter of the form $A\exp(-\alpha R)$ where R denotes the separation between oxygen atoms ($\alpha \geq 0$).

Thus, diminishing oxygen-oxygen separations during the c-axis contraction (note that the relative change $\Delta c / c$, in absolute value, exceeds the ratio $\Delta a / a$ more than twice) can cause the broadening of the oxide valence band and the corresponding narrowing of the film band gap.

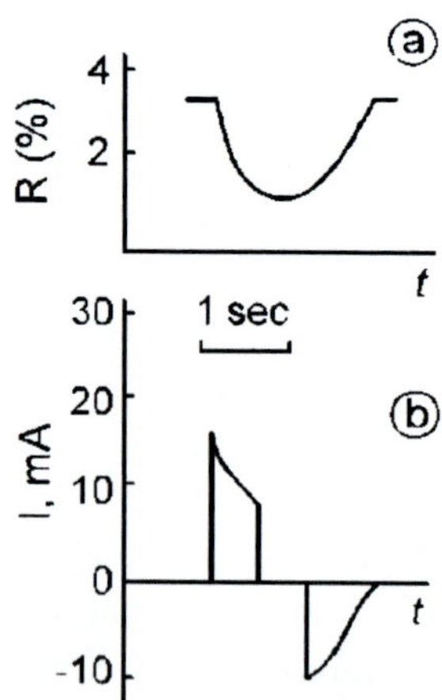

Figure 7. The time dependence of the intensity of the reflected light (a) and the current (b).

PART C

Currently, for applied purposes exceptionally is mostly used, despite of as an electrochromic material being considerably yielded to niobium pentoxide, revealing no degradation features at cycling and storage It is explained by the fact that until now ECE has only been observed at Nb_2O_5 polycrystalline layers, possessing apparent electrochromic properties in contrast to the amorphous films. At the same time, it is known, also based on the results of WO_3 analysis, that from the point of electrochromic process flow efficiency (ECP), polycrystalline films are hardly suitable for instrument applications, as long as they are featured by relaxation phenomena in ECP kinetics, caused due to the heterogeneity of polycrystalline structure.

Being so different, in terms of ECP, properties of Nb_2O_5 polycrystalline and amorphous films, bring us to the peculiarities of structure of these compounds, namely, to the surface morphology of the oxide layers, as long as it is known that volumetric properties of the amorphous structure are favorable for ECP. Comparison of amorphous and polycrystalline films of niobium pentoxide reveals significant differences in the morphology of the surface: developed micro-relief in case of polycrystalline oxide and highly-solid surface in case of the amorphous layers.

The question of the correlation between the oxide surface morphology and ECP efficiency can be resolved on the basis of the theory proposed by the authors [15] according to which dispersion of the system is considered as an active thermodynamic variable that can shift the direction of the process (including, in case of massive phase) ensuring its efficient progress. At the same time development of the surface, which increases the dispersion of the system, causes growth of concentration of point defects (vacancies), just as it takes place at temperature increase. In regards to ECP, this means that development of oxide surface, which leads to an increase in concentration of oxygen vacancies, being predominant type of defects in WO_3 and Nb_2O_5 [16], can not only intensify ECP flow, as noted by the authors [17], but also activate it at conditions at which it hasn't previously been observed.

In the present survey for the purpose of experimental verification of possibility of ECP activation in amorphous niobium pentoxide by changing of the morphology of oxide surface it has been developed a method of producing of amorphous Nb_2O_5 films possessing developed surface.

It is known [18] that the anodic oxide film (AOF) on aluminum can be produced with a porous surface in electrolytes which etch oxides. Thus, oxide layers produced in an electrolyte containing etching ingredient - hydrofluoric acid (HF) is considered as the key object of study in this work.

Niobium in the form of foil, annealed at the temperature of 2300K in vacuum, not less than 10^{-5} Torr, has been oxidised in electrolyte containing 25 ml of phosphoric acid (H_3PO_4) and 25 ml of HF per liter of water alternating asymmetric current with cathode half-cycle to anodic amplitude ratio in the amount of 5:1 in two stages: first at a fixed current density j = 5 mA/cm2 (galvanostatic mode) up to a voltage of 40-50 V, and then at a fixed voltage U = 10V (voltstatic mode) for 20-30 minutes.

The resulting films had a light gray uniform color. The oxide layer thickness was dependent on the oxidation mode, and ranged from 0.2 - 0.3 mm with a shutter speed within 1 hour at a voltage of 50 V. The films are amorphous, as evidenced by the halo electron diffraction obtained from them.

Figure 7 shows time dependencies between the intensity of the reflected light and current received for Nb_2O_5 AOP (coloring voltage was 1 V, area of the sample S = 0,5 cm$^{2)}$.

Thus, the applied method of niobium oxidation allows obtaining of amorphous films of Nb_2O_5 with apparent ECE, as a result of oxide surface development. The resulting oxide layers can withstand not less than 10^6 cycles of staining - bleaching, as well as long-term storage without any signs of degradation.

Given results support the decisive role in the ECE oxide surface. Detection of correlations between surface morphology and ECE efficiency contributes to optimization of Nb_2O_5 electrochromic amorphous films producing technique, by means of both anodic oxidation and vacuum condensation, which reveals prospects of niobium pentoxide application in electrochromic display devices.

An alternative was found to electrochrome process (ECP), and it is the process of recombination of hydrogen atoms emitted on the surface anode oxide film (AOF) and their subsequent evacuation as gas bubbles, which considerably limits the ECP rate. We have found that when Nb_2O_5 AOF is polarized by the packages of microsecond pulses, starting potential of gas emission can be enhanced three- to fivefold, which promotes significant increase in current density on film electrode, and, thus, raises the ECP rate.

A number of papers discuss the emergence of states differing in their life times and emergence nature on AOF surfaces. Thus, slow surface states (SSS) generated by the electrolyte chemosorbed ions participate in adsorption and even initiate it [19].

It is found that ECP can take place similarly to the catalytic mechanism of dissociated chemosorption. Kinetic factors of such mechanism can be taken into account upon the examination of diffusion equation and Frumkin's isotherm [20]. The equation first term contains Q parameter, characterizing the extent of surface filling during adsorption. It is possible that Q parameter, constant for normal conditions, can grow by the superposition of packages of microsecond-duration pulses, i.e. electroadsorption effect is realized. The Q growth accelerates ionization of hydrogen atoms and their subsequent diffusion to the AOF solid phase, and prevents the development of rivaling recombination and gas emission processes.

At analysis of the field effect on Ge, the authors [21] have determined that attachment of transverse electric field pulses to the sample causes positive charge accumulation in the surface states (related to water adsorption). Such accumulation is observed due to the fact that a positive charge is not capable to discharge within the time between pulses.

It has been shown that the pulses of transverse electric field, applied to the sample, result in the accumulation of positive charge in SSS associated with water adsorption, which has no time to resolve in the interpulse period. The "accumulation effect" was explained by electroadsorption. In our case, it is also evident that polarization of pulses with period-to-pulse duration ratio smaller than the SSS life times result in the increase of surface concentration of OH_3^+ - groups, caused by the accumulation effect, and in subsequent discharge of hydrogen atoms.

In other words, the application of pulse electrochemical polarization by the packages of microsecond pulses with low period-to-pulse duration ratio promotes significant increase of ECP rate in niobium pentoxide at the expense of added filling of AOF surface by hydrogen ions.

REFERENCES

[1]   Sieber, I., Hilderbrand, H., Friedrich, A., Schmuki, P. *Electrochem. Comm*, 2005, vol. 7, pp 97-100.

[2]   Choi, J., Lim, J. H., Lee, S. C., Chang, K. J., Kim, K. J., Cho, M. A. *Electrochim. Acta,* 2006, vol. 51, 5502-5507.

[3]   Jouve, J., Belcasem, Y., Sevsrac, C. *Thin Sol. Films*, 1986, vol. 139, pp 67-75.

[4]   Kofstad, P. *High Temperature Oxidation of Metals* J. Wiley & Son: New York, 1986, pp 56-73.

[5] Oechsner, H.; Giber, J.; Fuber, H.; Darlinski, A. *Thin Sol. Films*, 1985, vol. 125, 199-210.

[6] Umanskiy, Ya. S. *Rentgenographiya metallov i poluprovodnikov*, Metallurgizdat: Moskow, 1969, pp 277-289.

[7] Chen, W. K., Swalin, R. A. *J. Phys. Chem. Solids* 1966, vol. 27, pp 57–64.

[8] Gusinskiy, G. M., Muzdaba, V. M., Tomilenko, G. F., Khanin, S. D. *Z. Tech. Fiz.* 1986, vol. 56, pp 204–205.

[9] Kroger, F. *Chemistry of Imperfect Crystals,* Mir: Moskow, 1969, pp 602 –654.

[10] Myamlin V. A.; Gurevich, Yu. Ya., *Izv. Akad. Nauk SSSR, Ser. Khim.,* 1964, vol. 12, pp 2237–2241.

[11] Chazalviel, J.-N., *Electrochim. Acta*, 1988, vol. 33, pp 461.

[12] Milns, A. *Deep-Level Impurities in Semiconductors* Mir: Moskow, 1977, pp 479 -562.

[13] van Zandt, L, L., Honig, J. M., Goodenough, J. B. *J. Appl. Phys.*, 1968, vol. 39, pp 594 – 596.

[14] Gubskiy, A. L., Kovtun, A. P., Khanin, S. D. *Fiz. Tverd. Tela,* 1987, vol. 29, pp 1067-1071.

[15] Lidorenko, N. S., Schizik, S. P., Gladkikh, N, T. *DAN SSSR*, 1981, vol. 257, pp 1114- 1117.

[16] Kofstad, *P. Deviation from stoichiometry, diffusion and electroconductivity in simple metal oxides*, Mir: Moskow, 1975, pp 371-396.

[17] Pilipovich, V. A., Budkevich, B. A., Romanov, J. M., Ivlev, G. D., Ges, J. A., Zvavyi, J. A. *Phys. Stat. Sol (a)*, 1985, vol. 89, pp 709–719.

[18] Young, L. *Anodic Oxide Films,* L.-N.-Y.: Academic Press, pp 235–377.

[19] Kao, Kwan; Hwang, W. *Electric Transport in Solids*, Pergamon Press: Oxford, 1984, pp 327–350.

[20] Reichman D., Bard, A., Laser D. *J. Electrochem. Soc.*, 1980, vol. 127, pp 647–654.

[21] Rzanov, A. B.; Vlasov, Yu. F.; Neizvestniy, I. T. *Z. Tekh. Fiz.*, 1957, vol. 27, pp 2240-2250.

INDEX

D

E

O

P

Q

R